HIDDEN VALLEY OF THE SMOKIES

Ross E. Hutchins

HIDDEN VALLEY OF THE SMOKIES

With a Naturalist in the Great Smoky Mountains

ILLUSTRATED WITH PHOTOGRAPHS
BY THE AUTHOR

DODD, MEAD & COMPANY · NEW YORK

ACKNOWLEDGMENTS

Thanks are due to the personnel of the Naturalist's Office of the Great Smoky Mountains National Park for their fine cooperation. The author wishes also to thank the following people for their aid at various times: Supervisor Ranger H. R. Penny, Mr. George B. Stevenson, Mr. and Mrs. Allen Puckett, Mr. and Mrs. Ambrose Gaines, Mr. and Mrs. E. S. Metcalf, Mr. and Mrs. Carlyle Potter, Mr. and Mrs. Tom Morresy, and Mr. J. T. Higdon. Special appreciation is due my wife, Annie Laurie, for her help and encouragement in ways too numerous to mention.

To memories of a thousand campfires
burning brightly, and of the sounds of
tumbling waters in the night

Foreword

THE GREAT SMOKY MOUNTAINS are a mecca to both vacationers and nature lovers. In spring they gather here to marvel at the beauty and variety of the wild flowers; in autumn they return to view the spectacular display of colorful foliage. Even more visitors come in summer to enjoy the scenery and the mountains' coolness, and to feel, for a time, an intimacy with nature.

Hidden away in this vast expanse of mountains are secluded valleys and glens, most of them with bright streams tumbling down through the forests with pleasant, rippling sounds. Occasionally the streams drop over ledges, forming waterfalls of spectacular beauty. These rivulets unite with others along the way, forming rivers that continue down the valleys and out of the mountains.

One such valley lies cradled between high mountain ridges extending northwestward from Clingmans Dome, the highest point along the Great Smoky Mountain range. Down this forested valley flows a turbulent river, its size augmented, here and there, by small streams rushing down the mountainsides through shady vales, green with mosses and ferns. Farther down, the river passes through a deep, tortuous gorge, flowing away, at last, from the highlands that gave it birth. This is Little River.

The upper reaches of the valley are now inhabited only by the wild creatures of the mountains, living undisturbed lives in the lush, sylvan setting. Here a person, especially a lover of nature, may for a time, find peace and solitude away from the populous world. Hopefully, he may also learn something about himself. It is a place for the inquiring mind, a place to marvel at primitive beauty, seeing nature firsthand in a wild setting. This is the place I call Hidden Valley and I invite you to share my excursions through its shadowy forests and to listen with me to the music of its murmuring streams.

R. E. H.

Contents

Chapter 1

HIDDEN VALLEY

HIDDEN VALLEY is a place of great beauty and, to me, symbolizes peace and tranquility. Isolated, set apart from the crowded haunts of men, its forest trees stand tall and somber, rising above the most varied and lush vegetation imaginable. The earth is carpeted, here and there, with mosses, green and soft as velvet, and by delicate ferns in abundance. In spring the valley abounds with wild flowers, and in autumn its mountainsides suddenly blaze with spectacular color. Now and then a deer moves gracefully through the shadowy forest, then fades silently from sight, not a leaf disturbed by its passing. A river flows down the valley, rushing over great boulders and spilling into crystal pools. Small streams tumble down from the higher elevations along the river's course, dropping from ledge to ledge between green walls of laurel and rhododendron. The valley is a place of muted sound—the distant calls of birds, the continuous murmur of boisterous water, and the soft breath of the wind in the trees. This is my Hidden Valley.

The valley is situated deep within the Great Smoky Mountains National Park, which was created in 1934 and comprises about eight hundred square miles of mountainous forest. It is divided approximately through its center by the boundary between Ten-

Down from the heights in Hidden Valley flows Little River, spilling between great lichen-covered boulders.

1

nessee and North Carolina, which is also the central crest of the range. The topography of these mountains is complex in the extreme, cut into deep gorges and valleys, separated by high ridges in, seemingly, helter-skelter fashion. Naturally, the streams on the northern approaches to the mountains flow northward, while those on the southern approaches flow southward. However, the southward-flowing streams mostly curve westward and then northward just south of the mountains, uniting their waters eventually with those from the northern side, and forming the Tennessee River system which drains most of this vast area. To understand how this came about one must recall that these mountains were long ago thrust upward during the Appalachian Revolution, the streams being forced to follow their natural drainage pattern, cutting deep valleys and gorges.

As compared to the Rocky Mountains, these are pygmies; their highest point, Clingmans Dome, rises but 6,642 feet above the sea. Until I had reached the age of more than twenty years I had never been below 5,000 feet elevation. The ranch where I grew up in the Rockies lay in a deep valley whose elevation was 5,850 feet, with surrounding peaks towering to nearly 11,000 feet. Yet here, near the crest of the Great Smoky Mountains, at less than 5,000 feet, I look down across a jumbled mass of mountains that is almost frightening in its immensity.

The heights of mountains are deceiving. In the Rocky Mountains the entire land is at a high elevation; the bottoms of the valleys may be five or six thousand feet above the level of the sea, with peaks rising far above. Thus, the western mountains themselves "begin" at high elevation. The Great Smoky Mountains, by contrast, rise from near sea level, thrusting their heights to beyond 6,000 feet. In a manner of speaking, they are almost as high as the Rockies and nearly as massive in appearance. Yet everything is relative; I once heard a man in the Deep South call a 500-foot hill a "mountain," while on another occasion—this time in Wyoming— I heard an old cowhand, seated beside a campfire, remark that he had once "punched cows" over on the other side of the hill. He

Looking northeastward from Fighting Creek Gap, the Great Smoky Mountains stretch away, ridge after ridge, with deep valleys between. Often, in autumn, the valleys are filled with fog. At other times, a blue haze hangs over the mountains.

was pointing toward the Teton Mountains, towering more than 14,000 feet into the night sky!

The Rockies differ from these mountains in yet another way. Geologically, they are relatively young; their formation began "only" about 100 million years ago as contrasted to the Smokies' 500 million years. Thus, the Smokies are at least five times as old. These two large mountain masses had similar origins; both arose from geosynclines, or submerged seaways, and both ranges were once below the level of the sea. It is one thing to read such facts, but when one actually views the evidence firsthand, the impact is far greater. I once climbed to the top of a 10,000-foot peak in the Rockies and was astonished at seeing thousands of fossilized crinoids imbedded in the shaley stone. I knew these animals to have been inhabitants of the ancient seas, yet here they were, nearly two miles above their place of origin.

There are no fossils, plant or animal, in the higher parts of the Great Smoky Mountains. This is because these mountains were thrust upward from their ancient birthplace, the inland seaway, long before life began on earth. They are of Pre-Cambrian origin, among the world's oldest mountains and almost everything about them bespeaks venerable age. Their once-harsh contours have been worn down by rushing streams and their rocky skeleton weathered away, forming rich, porous soil where plant life in amazing diversity thrives in an equitable climate. I know of no other mountains, except perhaps in tropical regions, where the valleys and slopes are so densely clothed with lush vegetation, in many places making progress through it almost impossible. The vapors rising from this mass of living, breathing greenery are suffused with volatilized oils, imparting to it a blue haze that almost always hangs over the mountains, resulting in their name, Smoky Mountains. In the Cherokee tongue they were thus known, and early travelers followed suit in naming them.

It is not at all difficult to be lost in these mountains; there have been many instances of people, including adults, completely disappearing. As recently as 1969, a small boy was lost along the Appalachian Trail near the southern end of the Park and was never found, even though he had been wearing a bright red jacket. This, in spite of a massive search carried out over many weeks by thousands of volunteers both on the ground and in the air. To a person who has never hiked through these mountains this may seem incredible, but to one familiar with the steep, densely forested terrain, it is not at all difficult to understand. The ground is covered with spreading growths of dog-hobble, rhododendron, and laurel, as well as twining grapevines, often making progress impossible. Frequently the view is reduced to a few feet and a person soon finds himself entangled in a morass of creeping, climbing vegetation where all sense of direction is lost. It is easy to say that one has only to follow the old rule: Find a stream, and it will eventually lead you down and out of the mountains. This sounds fine, but what do you do when the stream sinks beneath masses of great,

Dense as a tropical jungle, vegetation grows rank and lush in an equitable climate with abundant rainfall.

moss-covered boulders? What happens to your sense of direction when you must detour around and through dense tangles or "hells" of dog-hobble and laurel and thick stands of rhododendrons, their contorted trunks completely blocking the way? In addition, the mountains are dissected in confusing directions and a trail that seems to be climbing may, in fact, be descending. I consider myself to be a fairly good woodsman, yet I have often been bewildered and uncertain of my way. To me it is not at all remarkable that many people have been lost in the mountains. What is really surprising is that even more do not disappear.

Botanists consider these mountains to have been the cradle of much of our eastern vegetation. Once they had risen from the inland sea, they were never submerged again. Too, they were never scoured by Pleistocene ice sheets as were areas farther north. Thus, surrounded by optimum conditions of temperature, moisture, and soil, and uninterrupted by climatic and geological setbacks, the

Afternoon sunlight slants through the Hidden Valley forest, bringing the trunks of large trees into sharp relief.

plant life of these mountains evolved in astonishing diversity, gradually spreading away in all directions. Some of it eventually reached southern Florida, while other kinds extended their ranges all the way to the Great Plains and into Canada.

Probably no area of comparable size in the world supports such varied and abundant plant growth. This is not only evident in the trees and shrubs but in the herbaceous plants as well. The first time I came to these mountains in early spring before a single leaf had appeared on the trees, I was surprised at finding ground-living flowers in abundance; almost everywhere, they were pushing up out of the moist earth. There were trilliums and violets and orchids, as well as masses of white phacelia and other flowers. The truth is, of course, that in the absence of tree leaves as a screen, the spring sun had warmed the ground, stimulating the growth of the ground-dwelling plants. Later, when the tree leaves had unfolded, the earth below was in shadow, protected from the increasing heat of the sun. This condition prevailed all summer, aiding the shade-loving plants to survive. In autumn the cycle continued; the leaves fell, allowing the sun to again reach the earth and to bathe the fall-blooming flowers in sunshine and stimulating the formation of seeds. It is a cycle as old as the mountains and illustrates nicely how well the seasonal development of the plant life fits into their eco-logical needs.

Not only has this area been a haven for the development of a remarkable collection of flowering plants, but for the growth of a varied host of so-called lower plants as well. Here grow more than fifty different kinds of ferns, probably more than in any other similar area. As might be expected in such a cool, damp climate, it is also a haven for large numbers of mosses and fungi. In late summer and autumn, mushrooms appear on the ground almost everywhere. Other kinds of fungi, strange in form and coloration, grow upon rotting logs. These primitive forms of plant life vary in color from bright red to vivid green. Truly, it is a place where nature runs riot.

My Hidden Valley is a sequestered area, cradled between high mountains. To the east lies Sugarland Mountain and to the west rises a series of mountains and ridges. I call it Hidden Valley with good reason; to me, that name is most descriptive of its nature. Places I love I usually designate by my own special names; it is like giving a dear friend a nickname, and thus named, a place becomes "mine." Still, in fairness, I will tell you the way to Hidden Valley that you, too, may share the pleasures of an easily accessible but most fascinating spot—easy, that is, if you view it only from the meandering road that follows up along the river. If, however, you are endowed with curiosity and the spirit of exploration, you may struggle through the dense vegetation, climb over or around great boulders, or push through thick growths of ferns and stinging nettles. If you have the stamina you may also climb the steep mountain slopes rising upward from the valley floor toward the high backbone of the range, seeking always to discover what lies beyond. This is the true lure of the mountains, the mystery of hidden places, of finding secluded spots, of venturing into unknown country where paths are dim and far between. You may follow trails or old logging roads, but that is the easy way. In any case, you will never know what lies beyond each boulder, over each ridge, or around the next bend in the trail unless you go and see. Perhaps it will be merely more jungle-like forest but, on the other hand, you may abruptly be confronted by a beautiful waterfall tumbling down over mossy rocks. You may also find a cluster of yellow lady's-slipper orchids, golden against the dark vegetation, or perhaps a group of strange Indian pipes, pale and ghostlike in the half-light of the forest. If you are lucky you may suddenly come upon a deer, half-hidden in the forest's dappled pattern of sunlight and shadow. Yet I must warn you, my venturesome friend, walk cautiously; you may also encounter a copperhead or a timber rattlesnake, graceful in form and coloration, but more dangerous than a bear. Mixed with the beauty of these mountains there may also be hazards.

A deer may pause to gaze at the human intruder with alert eyes, then fade silently away through the forest.

Hidden Valley lies in the heart of the Great Smoky Mountains. Little River gathers its waters from numerous small streams flowing down the northern watershed below the crest of the mountains.

To reach Hidden Valley from the south you must drive up the winding road through Little River gorge. This road follows along beside the tumbling waters of the river, rounding one sheer outcropping of stone after another, flanked on either side by steep mountainsides. The way is beautiful and spectacular and, having driven up the tortuous roadway, you will be surprised to learn that it was actually built on the roadbed of an old logging railway. If, however, you arrive from the north you must follow the highway through Fighting Creek Gap, then drop down into the valley of Little River. Beyond the gap the road eventually approaches the river, but at a designated point it turns left to Elkmont and then goes on up the valley. Several miles beyond this point you will pass the last summer cottage and suddenly come to the end of the surfaced road. From here on, the road is meandering and narrow, hemmed in on one side by great boulders or wall-like outcroppings of stone, the ancient skeleton of the mountains. On the opposite side is the stream, tumbling always over its uneven bed, rushing over the rocks and pouring here and there into deep clear pools. This is the vale I call Hidden Valley.

In truth, I consider the designation of this watercourse as a "river" to be a misnomer; somehow I picture a river as a wide stream flowing quietly down a valley or across a broad plain. Little River is actually a turbulent mountain stream; in the West we would call it a creek.

Little River gathers its waters from smaller streams dropping down the mountain slopes along its course; some are sizeable, others mere trickles. As you go up the valley you will pass a small brook tumbling down the side of the mountain from the right. It splashes over mossy rocks between fern-covered banks and beneath mossy logs that have fallen across it. This is Bear Wallow Branch. Beyond this point the roadway meanders up the valley, continuing along the right-hand side of the stream, passing, eventually, near a

Little streams cascade down the mountainsides over boulders green with moss. Brook lettuce, left, grows along the margins.

deep, dark pool into which the waters pour over great boulders with thundering sounds. Here you will no doubt pause for awhile, drinking in the beauty of the scene and listening to the sounds of the stream, wishing, perhaps, that you could remain forever. Here my wife once set up her easel and for several hours busily applied brushes to canvas. Then, glancing at the scene, she was astonished by the sight of a bear jumping from boulder to boulder across the stream.

Beyond the pool the narrow road leaves the river for a time, passing across an area of level forest, carpeted everywhere with ferns. Later, you will rejoin the stream, arriving at Hidden Valley's most spectacular and beautiful waterfall. As far as I know, this waterfall has no name; to the unimaginative mountaineers the stream was known merely as Huskey Branch. Its waters fall down the mountainside and into a deep pool of the river where trout may always

Here and there, the river pours into pools where trout may be seen in the crystal-clear water.

Bands of light and shadow ripple across the clear pools, creating reticulated patterns upon the bottom.

be, seen, moving slowly through reticulated patterns of light and shadow gliding slowly over the rocky bottom. A local friend calls this her "blue pool" and often swims there—she says—although the waters of the river are very cold, even in midsummer.

It is the falls of Huskey Branch, however, that I admire more than the stream itself. When seen from below, they drop over distant ledges high above, falling whitely into small basins in the rocks, then spilling over yet other ledges on their way down to the pool. Looking upward from the roadway, the stream seems to appear out of nowhere; yet I have explored its upper reaches for some distance and know that it flows down from the mountains, often rippling across fairly level little glens, making its way through seemingly impenetrable jungles of laurel and rhododendron.

Beyond the falls the road crosses the river and continues on up Hidden Valley, sometimes beside the stream, at other times deviating from it. At one place the main river is joined by Fish Camp Prong, a sizeable stream dropping down from the heights. Eventually, several miles up the valley, you will arrive at road's end—I

Tree squirrels are common in the valley, jumping through the trees or peering at the traveler from the safety of trunks.

call it "head of navigation"—and here you must either turn around or continue on foot. If you travel this road in summer while the trees are in foliage you will see few distant vistas; the crests of the far-off mountains are screened from view in all directions. Only in autumn or winter when the trees are bare are you likely to see great distances.

There are other ways to enter Hidden Valley. From the east you can climb to Huskey Gap and then drop down into the valley. Or you may follow the trail leading out of Jakes Creek, entering the valley via Cucumber Gap over an old logging road. This is my favorite hike. In any case, remain close to the trails unless you are accustomed to wilderness travel through unknown forests. While these mountains are situated close to many great centers of civilization, you will never feel more alone than when wandering through them. If, like myself, you thrive on sylvan beauty and a sense of seclusion, this is your domain. I have trod its trails in spring, summer, and autumn, always finding peace and contentment.

Fallen logs decay, returning their nutrients to the soil. Some of these trees were cut long ago to obtain bark for the making of tanbark.

Chapter 2

THE INCOMPARABLE FOREST

THE FOREST OF Little River Valley stretches almost unbroken from the high, cool ramparts of Clingmans Dome down to Elkmont where nearly three hundred people once lived. The logging railway then snaked up the gorge, penetrating far up the valley. Today the railway is gone, its bed now a meandering roadway; all that remains as proof of the railway's existence are a few rusting spikes protruding from rotting logs and stumps. The forest is slowly reclaiming its ancient domain, lapsing once more into peaceful solitude. The coming of man was but a small incident in the long story of the valley. To paraphrase a line from Robert Service, it is now a valley "unpeopled and still."

The original settlers, the railroad and timber workers, have disappeared, leaving behind only the stone foundations of their cabins surrounded by growths of English ivy, and a sizeable cemetery bordered by an encroaching jungle of rhododendrons and great forest trees. Still, there are some who remember, who now and then visit this hidden cemetery to place flowers on the graves of loved ones of the past. In visiting this burial place I have been impressed by the number of small children resting there, evidence of high infant mortality rate among the mountain people. There are indications, too, that once the crucial years of childhood had passed, the people often lived to great age.

I have come to know the upper reaches of the river as Hidden Valley, and to me this name is most apt. It is hidden away and the

17

The forest trees of Hidden Valley rise high above the earth, shading the lesser plants growing below. About 150 kinds of trees are native to the Great Smoky Mountains.

streams spread finger-like up the steep escarpments to the high backbone of the Great Smoky Mountains. Hidden Valley rests in Tennessee but beyond the crest of the mountains lies North Carolina. The valley's higher slopes are covered with dense stands of red spruce and Fraser fir, interspersed with growths of mountain ash and other trees and shrubs. The truth is that the vegetation on these higher elevations is alien in nature, resembling closely that found a thousand miles to the north. Seemingly, these upland forests are leftovers from the Ice Ages that long ago pushed cooler climates and more northern vegetation southward. This alpine vegetation covers the roof of the mountains like a dark green blanket as they rise into the blue, misty haze that almost always covers them except for brief periods after rains have washed the air. Only then are the more distant vistas seen. Deeper down, the valley is frequently bathed in dense fog, especially in autumn; it hangs like lazy, white clouds, spreading through the forest, condensing upon the leaves, the droplets gleaming upon the ferns and mosses carpeting the valley floor.

It is a region of high rainfall, the yearly average in the valley being more than sixty inches. Fortunately this moisture is rather evenly distributed through the year, about five inches falling during most months except autumn when the average drops to a little over three inches per month. In winter, precipitation is frequently in the form of snow, which often blankets the earth to depths of two or more feet. With respect to temperature, the January average is about forty degrees above zero, but temperatures as low as minus twenty degrees have been recorded. The July average is 72.6 degrees, but it often becomes quite hot, 98 degrees having been noted. According to long-term averages, the first killing frost of autumn occurs on October 17, while the last spring frost comes on April 24. As might be expected, there is about a week's difference between the first and last frosts along Little River (elevation 2,500

The misty road through Hidden Valley. Often, in autumn, clouds hang low in the valley, bathing roadway and forest in mist. The ground at that season is carpeted with fallen leaves, golden and scarlet.

Ferns grow everywhere; they carpet the floor of the valley with their filmy leaves and take root upon rocks and fallen logs. A dead birch has fallen among these ferns.

feet) as compared to Gatlinburg, a thriving city on the other side of Fighting Creek Gap, at an elevation of about 1,500 feet. The first frost occurs a week later at Gatlinburg, while the last spring frost comes, on an average, a week earlier there. Also, Gatlinburg receives about ten inches less rainfall per year than do the upper reaches of Little River. In general, the higher elevations of the Great Smoky Mountains receive more moisture than the lower areas, in some places amounting to more than ninety inches per year!

This abundant moisture, falling upon the mountains, flows down the slopes, gathering, here and there, into numerous streams that tumble down through mossy glens, dropping from ledge to ledge among dense growths of ferns and other moisture-loving plants. Immediately after heavy rains the streams quickly rise, carrying away the excess water, then resume normal flow again, crystal-clear and cold, fed by a constant supply of moisture percolating slowly through the porous soil of the mountains.

All of the climatic factors add up to ideal conditions for lush plant growth, conditions that have prevailed for thousands of years and stimulated the growth of one of the world's most remarkable stands of great trees. Here, there once grew great tulip poplar trees as well as pines and spruces. Scattered through the forest of Little River Valley may still be seen great stumps, all that remain of forest giants that once rose, straight as arrows, above the surrounding timber. Often one sees large trees growing out of the decaying stumps of once-great trees. In one place, I saw a stump upon a stump with a small tree growing out of the second one. This is a place where shrubs refuse to remain shrubs; often they grow to tree size. Buckeyes, which in most places are regarded as shrubs or at least small trees, here have been known to reach 125 feet. Mountain laurel, usually a shrub, may grow to tremendous size, one specimen measuring more than six feet through. Poplars often reach 200 feet and hemlocks 100 feet. There once grew in the valley great chestnuts, their trunks measuring nearly eleven feet in diameter, but they are gone—destroyed, first, by the loggers' axes and, later, by chestnut blight.

Here and there in the mountains one often sees large chestnut logs and stumps, now moss-covered and slowly rotting away. In a few places one finds sprouts growing beside these decaying stumps, pitiful efforts of the trees to survive. But these sprouts never be-

In autumn, daddy longlegs spiders emerge at night from hiding and walk across the ground or over the low vegetation of the humid forest.

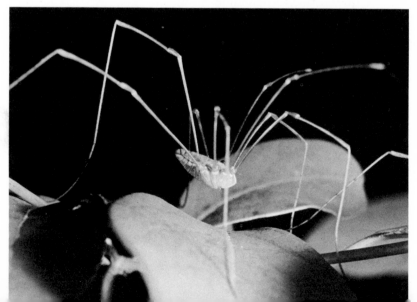

come more than sprouts; they soon die, victims of the blight disease. This disease, a fungus native to the Orient, first appeared in a zoological park of New York City, probably having been carried into the country on nursery stock. Ten years later it was found in the vicinity of the Smoky Mountains and has now destroyed every chestnut from Canada to the Gulf, certainly an American tragedy of major proportions. All the wild chestnut trees are gone.

The passing of the chestnut trees has had a profound influence on the ecology of the mountains. In former years chestnuts formed an important part of the food of both bears and gray squirrels. Acorns were also eaten, but to a lesser extent. With the disappearance of the chestnuts these animals were forced to feed largely upon acorns. However, for some unknown reason, there are often years when the acorn crop fails. The result has been that during such times bears leave the sanctuary of the National Park, moving into surrounding areas where they are hunted. Gray squirrels, once very abundant, are now rarely seen.

Almost always alone, I stroll through the stately forest feeling as if I were in a great cathedral. The songs of birds, like the sound of a distant choir, filter down from far above. I find myself treading softly along the dim trails, fearful of breaking the spell.

I often wonder how these mountains compare to those seen by the early observers and travelers. Certainly the primitive forests, untouched by the axe, must have been spectacular and beautiful. On the other hand, the vicissitudes of travel no doubt detracted from the appreciation of the beauty of the land through which the early explorers passed. Probably the first Europeans to visit these mountains were Hernando de Soto and his men who came seeking treasure. This expedition was financed by six hundred gentlemen of Spain and Portugal in the nebulous hope of receiving great riches from their investment. De Soto came but forty-eight years after Columbus had discovered the New World, viewing the untouched wilderness with eyes that sought only gold. The beauty of the land no doubt escaped him; he saw only savages in a strange country and dense vegetation that made travel almost impossible. He did visit the Cherokees—he called them Chalaque—and so it seems probable that he may have passed across the southeastern foothills of

At one time great chestnuts grew in the forest, but all that remains of them are decaying logs covered with mosses.

Plants and trees often grow out of cracks in the large boulders. A blooming dog-hobble thrives in a cleft in the stone.

the Great Smoky Mountains. Finding only hardships and difficulties instead of treasure, de Soto tortured and enslaved the Indians, rendering the travels of later explorers more hazardous.

After de Soto, there ensued a long period when this mountain area remained free of European exploration or settlement and, for a long while, nothing much was added to the knowledge of the area regarding its plant and animal life, or of the topography of the mountains. There were, indeed, a few travelers, but no naturalists having a true appreciation of the flora of these great mountains.

In the year 1699, only ninety-two years after the establishment of the first English colony, there was born John Bartram, the first American naturalist. During his life he collected and shipped to the British Museum hundreds of plant specimens, some of which are still preserved there. His son, William Bartram, was born forty years later and continued in his father's footsteps as a collector and student of plants. Imagine, if you can, the opportunity for exploration in the vast new land, as yet untouched by a naturalist's hand. It was prophetic, too, that even at that time William Bartram held the modern philosophy of conservation of natural resources through wise use. He had long dreamed of explorations in the South, a trip, as originally planned, that was to have taken two years. Actually he was gone for five, during which his wanderings took him all the way to the southern tip of Florida, across Georgia, Alabama, and on to southwestern Mississippi. Leaving Charleston in April of 1775, he explored the Old Charleston-Cherokee Trail to Cowee and the Little Tennessee River. During his five-year trip in the South, Bartram discovered numerous plants that were new to science and made many other valuable observations on the natural history of the region.

While Bartram never reached the higher elevations of the Great Smoky Mountains, his travels did take him into their southeastern highlands, and to anyone interested in firsthand impressions of the virgin forests of that time his journals make fascinating reading. One cannot help but be impressed by the extreme difficulties of travel through the unknown forests, of the manner of life of the Indians he visited, and by his continuing enthusiasm for biological studies under the most adverse conditions imaginable. However, in spite of the difficulties, I find myself envying this early traveler his

Clumps of dainty bluets thrive in pockets of earth between the roots of trees.

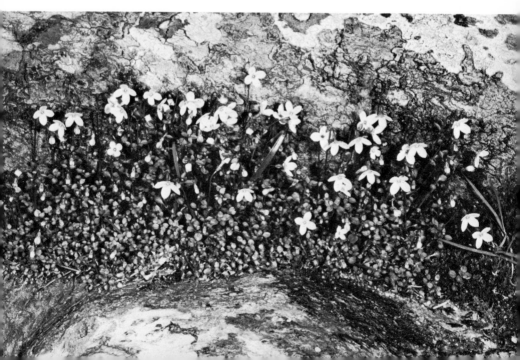

experiences in the wilderness area unchanged by the activities of man.

Probably the next botanist to visit the area was André Michaux, who had been commissioned by His Majesty Louis XVI of France to collect the most attractive flowers and trees of the New World for eventual transplanting at Versailles. He came to Charleston, South Carolina, in 1785, accompanied by his son, then seventeen years of age. Unusual among the early European collectors, Michaux brought several Old World plants as gifts. One of these was the mimosa, now one of the favorite flowering trees of the South. Apparently M. Michaux ventured farther into the wilderness than had Bartram. In his diary he, too, recorded the difficulties of travel as well as descriptions of the remarkable land through which he passed.

These early botanists were, indeed, privileged to see the mountains at their best, at least as far as their primeval splendor is concerned. They could not, however, have foreseen the changes that would come—the settlers, the hunters, and the loggers. Yet, basically, the mountains have been altered but little, and through wise management by Park officials the wild will slowly return to its own. In places such as Hidden Valley one may still feel the thrill of firsthand exploration, of tracing dim trails through the mountains, and of vicariously following in the footsteps of de Soto, Bartram, or Michaux. Often I have followed trails through the valley forest, seeing little to remind me that men had passed that way before.

Before the original stand of great trees was logged off, the forest floor was probably more open than at present and, as a consequence, travel less difficult. The cutting of the larger trees changed the forest ecology. There remain a few places in the Great Smoky Mountains, however, where the woodlands still stand in more or less their original grandeur, the great climax forest shading out the lesser growth far below. There, one may walk freely through the half-light, tinted emerald green by its passage through the foliage spreading high above. One's feet tread softly upon deep layers of

Sweet shrubs, with their deep maroon flowers, grow in many places in the forest. In autumn, they produce large pods filled with seeds believed to be poisonous.

ancient mold, making no sound. This is the way I picture the forests through which the first travelers passed. Hopefully, these conditions will return and history repeat itself.

Somehow, as I stroll through the forest of Hidden Valley, I always seem conscious of the past; perhaps it is because the past is so important to an understanding and appreciation of the present. To me the present is fleeting, only a brief moment in the vast span of time, and I constantly try to understand how it all came into being, how the great mass of leafy vegetation succeeded in clothing the primordial stone. Restless forces of growth are everywhere active and competitive. A plant—or animal—that cannot change and adapt itself eventually "goes out of style" and is replaced by a more aggressive kind. This, of course, is merely another way of stating Darwin's criterion of survival of the fittest. No plant that remains static can survive for long in the keen competition it encounters.

It is, of course, the trees of these mountains and valleys that take precedence over all other vegetation. Viewed from the air or from higher elevations in summer, the forests appear as an almost unbroken upholstery of varying shades of green, clothing the vast expanse of the mountains, dissected here and there by deep, dark valleys. It is an undulating sea of trees, frozen into immobility, the crests of its waves the high, cool upland ridges. When viewed from afar, these ridges rise tier beyond tier against the sky, each one a lighter shade of blue than the one preceding it.

Hidden Valley is but a small part of this mass of mountainous terrain, yet within it are contained almost every kind of tree and plant common to the area. To an amateur botanist, such as myself, this vegetation is overwhelming in its bewildering diversity. There are nearly 150 different kinds of trees alone, and more than this number of shrubs and vines. This is to say nothing of the herbaceous plants growing upon the forest floor. Deep in the valley the appearance of the forest is almost tropical. As one climbs upward toward Clingmans Dome the mountainsides gradually take on a far different appearance; there, one enters the Canadian Zone where the vegetation resembles that of central Quebec. It is mostly the elevation that makes this difference; climbing 1,000 feet is roughly equivalent to traveling 300 miles northward. Still, due to the complexities of prevailing winds and of moisture distribution, the vegetation refuses to conform to climatic patterns; often trees characteristic of the higher elevations mingle with semitropical types down in the valley. One of the most common trees in the area is hemlock, a conifer growing all the way from the valley floor up to 5,000 feet. At the lower levels it mingles with the pawpaw, a small flowering tree with strange fruit, most of whose relatives live in the tropics. Here, too, are found magnolias of three kinds. The Fraser or mountain magnolia, its large leaves lobed at their bases, grows up to 5,000 feet. There is also the umbrella magnolia with leaves nearly two feet long, the trees often growing forty feet tall. This tree reaches only up to the 2,000-foot elevation. In addition, there is the cucumber tree, not very common and with smaller

Left: *In summer, the white bell-like blooms of sourwoods hang from the twigs, turning upward after the flower tubes fall.*

Below: *In spring the deep purple blooms of pawpaw trees open along the twigs.*

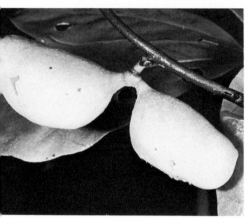

Above: *Maturing in summer is the strange-looking but edible fruit of the pawpaw.*

Left: *Among the most impressive of all the local trees is the tulip poplar, which belongs to the magnolia family. In spring, it bears large attractive flowers with orange throats.*

leaves than either of the other two. It may be encountered up to 5,000 feet. Little River Valley lies roughly at 2,500 feet and, as a result, the most common kind found there is the Fraser magnolia with its large leaves and ten-inch, greenish-white blooms. It often grows nearly fifty feet tall and individual specimens have been found measuring more than seven feet in circumference. One usually thinks of magnolias as inhabitants of warm, southern climates, yet here in this arboreal mixing bowl they mingle with trees of colder climates.

Among the largest of all trees in these mountains is another member of the magnolia family, the tulip poplar. Its botanical name is *Liriodendron tulipifera*, meaning a "tulip-bearing lily tree," which is quite descriptive of this attractive inhabitant of the forest. After spring hailstorms I have seen the ground beneath poplars carpeted with hundreds of the blooms knocked down by falling globules of ice. Hailstorms are common during spring in these mountains where violent weather is frequent. Falling hail beats off the leaves of the trees and tears the foliage of ground-living plants to tatters.

This birch grew from a seed dropped upon a stump, then sent roots down into the soil, so that it seems to be "sitting" upon the ancient stump.

Ambient temperatures may fall within a few minutes from eighty degrees to near freezing. Known locally as yellow poplars, these trees once grew to great size, often to more than a hundred feet tall, as is attested to by the large stumps remaining here and there in the valley from the old logging operations. Some of these stumps measure six or more feet across and some have second generations of large trees growing out of them.

In the humid climate of the valley it is not unusual to see trees of several kinds growing out of old stumps. I have noticed this to be a common habit of the birch, a habit that I presume gives the seedling tree a head start by lifting it above the ground where competition for space is less acute. Having started life upon the decaying stump, the birch then sends roots down to the ground and is safely on its way to continued growth. I am reminded of the similarity of this manner of germination to that of the tropical banyan or strangler fig. The seeds of these latter trees lodge in crevices in the bark of other trees and, upon germination, long roots snake down to the earth. Once anchored in the soil, the host tree is apt to die, strangled by the entwining roots of the banyan. Here in the valley the habit of beginning life upon decaying stumps is also found in rhododendrons, further evidence of the keen competition for space, of the restless forces striving and pushing upward to the sun. But the rhododendrons grow almost everywhere, their crooked stems and trunks twisted into fantastic shapes by vagaries of soil and rock and light.

One of the most attractive of all the large trees of the valley, especially from the standpoint of its flowers, is the mountain silverbell, called "peawood" by the settlers, a tree of the Styrax family, most of whose members grow in tropical regions, especially in South America. In spring, rows of white, bell-like blooms hang suspended from its twigs, replaced in autumn by large winged seeds. It often grows to great size; specimens more than a hundred feet tall and measuring nearly twelve feet in circumference have been found in the mountains. More common, however, are smaller silverbell trees with their attractive blooms within easy viewing distance.

One of the most attractive of all the trees of the valley is the mountain silverbell with its rows of bell-like blooms.

Attractive cones are borne in large numbers in the twigs of the hemlocks.

Almost everywhere in the valley and upon the mountainsides grow eastern hemlocks, their feathery foliage dark green and attractive. Unlike the pines, their needles are placed in rows on either side of the twigs, giving them a flat appearance. These conifers often are found beside the streams, with their "feet" in the flowing water. Seemingly, they can tolerate the shade of the deeper forest as well as the direct sunlight of the open areas. At the tips of their twigs they bear attractive little cones that, in time, fall to the ground, often in large numbers. This is not one of the great trees—it reaches but eighty feet—but it is a characteristic and attractive feature of the local forest.

It is not my intention to catalog all the trees of Hidden Valley; there are far too many. However, a few more should be mentioned because of their special interest in one way or another.

Easily identified is white pine. Its needles are in bundles of five and its limbs are in whorls around the trunks at various levels, usually five limbs at each level. It has elongate cones, averaging about six inches in length. This tree is common in the valley forest and often grows to more than a hundred feet tall. Like most of the other local conifers, it is a northern species, ranging southward from Canada.

Higher up on the slopes beyond the 3,500-foot elevation are found two other conifers that are relics or carry-overs from the cold of the Ice Age. The first of these is the balsam or Fraser fir with its flat, blunt-tipped needles, mostly arranged in one plane—that is, a row on either side of the twig as in the case of the hemlock. These are handsome trees of medium size, though they have been known to reach more than six feet in circumference. They grow from the highest elevations down to about 4,000 feet. The bark is thin, with numbers of large blisters filled with clear rosin. Early settlers imagined these blisters to be filled with rosin "milk," and so called them "she-balsams."

Another conifer of the higher elevations is the red spruce, identified by its four-cornered needles growing out of the twigs in all directions. The tips of these needles are spiny. These trees grow to

Maples growing side by side often graft themselves together.

larger size than the firs, sometimes to a hundred feet, and fourteen feet in circumference. The bark has no rosin blisters and, as a result, the settlers, by a strange twist of imagination, called these trees "he-balsams." Since these two kinds of conifers often grow in mixed stands, they evidently assumed that one, the fir, was the female while the other, the spruce, was the male. By peculiar quirks and misunderstandings do plants receive their local names.

Usually, when we think of maple sugar we automatically think of Vermont, but prime specimens of sugar maples have always grown in the Smoky Mountain area. The Park headquarters is located at Sugarlands, once a thriving little community whose chief industry was the production of maple sugar gathered from local maple trees. The sap was collected in buckets from spouts inserted in the trees, about four pounds of sugar, or one gallon of syrup, resulting from the boiling down of forty gallons of the sap.

The sugar maple (*Acer saccharum*), also known as hard or rock maple, grows to eighty feet and to three feet in diameter. Pyramidal or rounded in form, it thrives in the rich soil of the valley and on the hillsides. In summer these trees are attractive features of the forest, but to really appreciate them you must see them in autumn when they almost literally burst into flame with living color. Scattered here and there among the somber-hued conifers of the mountainsides these trees glow like columns of flame. Slowly, as the days pass, they take on more and more color, reaching their chromatic climax about mid-October, their coloration fading away as the leaves drop to the ground.

One of my favorite trees is the sycamore. While these do not usually grow to imposing size in the valley, they do sometimes reach large size in other places in the mountains, often as tall as nearly two hundred feet. They grow up to the mid-altitudes, that is, to the 3,000-foot level. To me, the sycamore is a beautiful tree because of its creamy-white bark, mottled by large, platelike scales

In a tree hole filled with soil, this fern found a place to grow high above the ground.

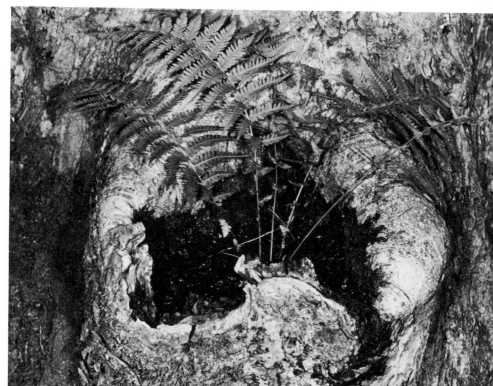

of irregular form. The sun, shining down through the foliage upon the dappled limbs and trunk, gives a most pleasing effect. Here and there on the trunks large limbs often decay, leaving gaping cavities or tree holes that became filled with water and in which dwell myriads of mosquito wigglers and other aquatic creatures. These holes are arboreal pools, isolated from other bodies of water. Too, sycamores have the habit of twisted, gnarled growth, often assuming strange, contorted, untreelike forms. Frequently they grow along the river, the large roots seeking anchorage among the half-submerged boulders. These roots seem to "flow" among the boulders like some thick, viscous fluid, dividing and uniting again and again in interlacing networks, binding the great rocks together in their tight embrace. Often these boulders are so completely enclosed by the groping roots that their dislodgement is impossible. In this way sycamore trees serve a useful function by holding together the soil and the rocks along the river's margin, thereby preventing the banks from eroding away during times of high water.

Along the river's margin the roots of a sycamore seem to "flow" over and among the boulders, as if groping for anchorage.

By some strange freak of growth, a large opening developed in the trunk of a sycamore beside the river.

I remember with special regard the sycamores of Hidden Valley, their gnarled forms reflected in the river's quiet pools where their masses of pink rootlets undulate gently in the crystal waters. I remember, too, the sycamores at evening, their graceful forms back-lighted in the rays of the descending sun. To me, somehow, these attractive trees symbolize the persistent growth forces of the forest. Perhaps it is because many of them, their great trunks half-rotted away, refuse to die; they continue to live in spite of decay and damage by wind and storm.

Sycamore roots spread across the ground, penetrating deeply into the earth. Among them grow violets and other plants.

Chapter 3

A WALK IN SPRING

SPRING IN HIDDEN VALLEY is a time of abrupt and radical change, the sudden transition from winter to summer. In early April the forests are, for the most part, still bare of foliage except for the evergreen conifers, the hollies, laurels, and rhododendrons. Mosses are green upon fallen logs and exposed boulders, while a few ferns, in sheltered places, begin uncoiling their fiddleheads. Robins sing their morning songs, reflecting their exuberance and joy of living; after each daytime shower they raise their "cheer-up" voices in the clean-washed air of the mountains. Migratory birds are now passing through the valley, pausing only long enough to search for food among the winter-bare trees, then flying on again, urged northward to distant nesting grounds by physiological changes in their small bodies brought about by the lengthening of the days. The warblers flash through the forest, bright spots of yellow against the somber hues of the hemlocks. Down the infrequent trails flutter spring's first tiger swallowtails, freshly emerged from chrysalids and now searching for flowers but finding few.

By mid-April many flowers begin opening in the valley forest and among the great masses of tumbled boulders along the margins of the valley floor. Dogwoods begin showing greenish-white,

A member of the lily family, the fairy wand grows in rich woodlands. It is also known as devil's-bit.

but their leaves are still folded in protective buds. Silverbell is just coming into flower, its rows of attractive, bell-like blooms white against the dark mountainsides. This early in the spring few other trees show any sign of reviving life; the forest still retains its winter aspect. As the lengthening days pass and showers fall, the trees, with seeming reluctance, begin unfolding soft green leaves and the forest gradually takes on a new appearance. Each twig becomes touched with green.

Probably my favorite spot along Little River is a place where its crystal-clear waters pour down among enormous boulders. Near the left side of the stream, at this point, the water spills down through a narrow cleft between two great stones, falling then upon a large, flat rock and spreading over it like a watery tablecloth. Around the edges of this flat rock the rushing sheet of water falls away like a white, lacy border, completing the illusion. Immediately below, the waters of the river gather once more, merging in a great, translucent pool where, this morning, I see the forms of trout moving indistinctly through the depths. On its downstream side the waters of the pool are obstructed by an enormous boulder, its contours rounded by centuries of rushing torrents. The water cascades around this boulder, but just above it the surface is marked by a series of ripple rings formed by the moving current. From my vantage point high above the pool I look down upon the panorama spread out below. It is a place of spectacular, eye-catching beauty and I always stop here for a time on my trips up the valley to drink in its charms. It appeals not only to my eyes but to my ears as well; all other sounds are obscured by the roar of the water rushing among the house-sized boulders jumbled together as if tossed there by some giant hand.

Today, a dogwood, now in full flower, stretches out over the pool, its white blooms reflected in the moving surface. In past autumns, this same dogwood was brilliant red, imparting to the pool reflected patterns of bright color.

Reluctantly, I turn my back on the lovely pool and cross the road, walking beside a small brook flowing down from the nearby

In spring, dogwoods without number open their snow-white blooms. Later, the foliage appears and, in autumn, their clusters of berries turn bright red.

mountain. It ripples over a shallow, sandy bed, its waters obstructed here and there by mossy stones and fallen logs. On some of the logs there are ferns, green and dew-covered, and on one a great millipede as large as my finger crawls slowly across the moss carpet. Hanging over the brook in one place is a dog-hobble, its racemes of white flowers just opening.

The sun rarely strikes this damp area, with the result that the vegetation here is green and luxuriant. Plants of almost infinite variety are now pushing up out of the dark earth and as I step from mossy stone to mossy stone I search for early spring flowers. Yellow violets star the green surface and, on one boulder, is a white violet, its roots having found anchorage in a small pocket of soil.

Following on along the brook, I am forced to climb over a large fallen log beyond which stretches a flat, marshy area. Here, I am

The blooms of purple fringed orchids, found at the tops of tall stalks, are lilac-purple with ragged lower lips.

pleased to discover half a dozen beautiful specimens of the purple fringed orchid (*Habenaria fimbriata*), its lilac-purple blooms clustered at the tops of tall stalks. The lower lip of each attractive, three-parted flower is deeply fringed. The finding of these flowers at this season is unusual, since their normal time of blooming is early summer.

But spring in these mountains is orchid time, the season when most kinds come into flower. Some bloom much later; the attractive yellow fringed orchid (*Habenaria ciliaris*) appears in midsummer. It grows in dryer places such as open woods, in contrast to the purple fringed orchid I find here in this marshy place.

A few feet beyond the brook and somewhat higher on the mountainside is another kind of plant in full bloom, one that I am always pleased to discover even though it is quite common in many places in the valley. This is wild ginger (*Asarum canadense*), a most unusual plant. Its leaves are kidney-shaped and from three to six inches in diameter, usually located close to the ground. Those not familiar with the plant may often overlook the blooms, since they are hidden beneath the leaves, frequently lying on the ground. Always intrigued by these strange flowers, I push aside a leaf and lift one up in my fingers. It is brownish-purple in hue, but its most peculiar feature is its shape; its calyx is cuplike and has three slender prongs extending outward from its edge. The blooms rest upon the earth or just above it, and I am at a loss as to what insects pollinate them; perhaps they cater to crawling insects or to small flies. Certainly its obscure coloration cannot attract bees or butterflies. Yet another unique feature of this strange plant is its roots. When I break one off and bring it close to my nose I am aware of a very pungent oder, the reason it is known as wild ginger. Early settlers concocted a brew from these roots that was believed to be efficacious in the treatment of colic and certain other types of stomach distress. Over in the next valley, along Fighting Creek, there grows another member of the same family. This is the little brown jug (*Hexastylis*), each plant with small, brown, juglike flowers near the ground.

Both wild ginger and brown jugs belong to the birthwort family (Asaraceae) and, on this morning, I am pleased to discover one of its relatives only a short distance farther up the mountainside. This is a Dutchman's-pipe vine (*Aristolochia macrophylla*), just as unusual in its way as its relatives. The vine twines up a small silverbell tree, and beneath one of its heart-shaped leaves I find one

The odd, three-pronged blooms of wild ginger are brownish-purple and located near the ground behind the kidney-shaped leaves.

Opposite: *Closely related to wild ginger is the little brown jug, with its characteristic, juglike flowers at the base of the plants.*

Dutchman's-pipe vines climb up the trunks of tall trees. The pipe-shaped blooms appear in spring.

of its oddly formed blooms. It is brownish-purple and resembles somewhat the form of an S-shaped pipe. Surrounding the boll of the "pipe" is an expanded flange of mottled color. The narrow throat of the flower is yellow.

Dutchman's-pipe flowers are unusual, but their pollination is certainly unique. Instead of perfume attractive to sweet-loving bees and butterflies, these strange blossoms are endowed with an odor alluring to small flies. These flies, enticed into a bloom, find themselves trapped for a time in its expanded base. Here they buzz about, becoming well dusted with pollen. Eventually they escape but are enticed into other Dutchman's-pipe flowers where their adhering pollen is rubbed off upon the flowers' stigmas, thus completing the process of cross-pollination. The Dutchman's-pipes of the Smoky Mountains are closely related to the much larger aristolochias of tropical lands. Some of the latter are very colorful and I remember with pleasure my past experiences in photographing them in their native habitats.

Wandering on up along the brook, I pass large boulders, moss-covered, resting upon the ground. They remind me of sleeping giants reclining quietly, half-buried in the dark earth. Here and there, on their mossy blankets grow violets and other small plants, while over them trail smilax vines, their leaves new and tender. In one place the brook spills over a broad ledge near the middle of which is a small moss-covered stone. Upon the stone's center grows a violet, its white bloom gleaming with droplets splashed up from the tumbling water of the stream. Adjacent to the stream, on the opposite bank grows a tall foamflower. One of its leaves just touches the rapidly flowing water. The leaf is swept downward, then springs back again, repeating the movement over and over in a sort of perpetual motion.

Foamflowers are common, especially where the soil is rich and damp.

May apple leaves push up through the earth. They will expand into an umbrella-like form, with white flowers beneath.

Two days later I returned and found, to my surprise, that the foamflower was still dipping its leaf into the stream. A simple calculation showed that it had done so approximately 100,000 times during the two intervening days, a tribute to the leaf petiole's resilience in withstanding the constant movements.

I am sure that I will long remember that little tableau, the foamflower leaf continually dipping into the waters of the brook, then springing upward again. It was a small, unimportant thing, yet it added its touch of motion to that of the rushing stream.

Everywhere along the stream plants of great variety are pushing up through the black soil. Virginia creepers are just coming into leaf; their leaves rise here and there on short stems, looking like tiny parasols. Later, these plants will climb upon the surrounding trees, draping them with their numerous five-parted leaves. Small white violets star the moist earth, and beyond the stream is a large patch of May apple plants, their green umbrellas rising nearly a foot on sturdy stalks. In time, white waxy flowers will open beneath each spreading leaf and, eventually, the blooms will be replaced by green lemon-shaped fruit filled with seeds. May apples (*Podophyllum peltatum*) are often known as "mandrakes" in these mountains, and both leaves and roots are poisonous. Also growing along the stream is brook lettuce, known as "lettuce saxifrage" (*Micranthes micranthidifolia*), a very common plant that thrives in wet places.

Almost everywhere I look, spring flowers are literally bursting out of the ground, stimulated by the warmth of the sun and by yesterday's shower. Wild strawberry plants are all about, a few in bloom. Later, their red fruit will appear, tasting far more delicious than that purchased in local markets. Yellow cinquefoil blooms rise above the grasses and the mosses, their delicate petals spread widely to attract pollinating insects.

There are abundant evidences of spring. The keynote seems to be lushness, of Nature running riot, stimulated by favorable amounts of moisture and proper temperatures. Seemingly the plant life of the valley is uninhibited by any check on rapid development; so fast does the vegetation develop that I can almost see it grow. Certainly, there are significant changes from day to day. The leaves of the oaks are now expanding, suffused by the presence of rust-red anthocyanin pigment that will later disappear. Upon the mountainside the graceful forms of fern fiddleheads rise above the ground like green question marks.

Along the stream's mossy edge is a dense growth of jewelweed and I am gratified to be able to recognize it at this stage of its growth. Most of the as yet flowerless plants I find difficult to identify; occasionally there is a familiar leaf, one that I know with certainty. The vegetation of these mountains is so lush and varied that only a professional botanist can know them all without their flowers.

As spring turns to summer the jewelweed blooms will appear, each one suspended below a leaf on a threadlike stem. There are two kinds; one has mottled-orange flowers, while the other is

The unusual blooms of jewelweeds tremble from suspensions beneath the leaves. Their explosive seed pods will appear later in the summer.

yellow. The seed pods, appearing in late summer, will burst at a touch, tossing the seeds a considerable distance. For this reason the plants are also known as "touch-me-nots."

Always of interest because of their unusual blooms are the jack-in-the-pulpits (*Arisema*). They, too, grow along the little brook, and I bend down to examine one of their green, vaselike flowers. It does not look like a flower at all, yet within its green spathe are enclosed all the parts of a complete flower having both male and female organs. By autumn each stalk will bear at its top a large cluster of bright red fruit. I have photographed these plants perhaps a hundred times, yet I can never resist the temptation of just one more picture. So I set up my camera and record this one on film, hoping, perhaps, that this photograph will be better than others already in my files. There are actually three different kinds of jack-in-the-pulpits found in the mountains, two of which have three-parted leaves. The other species (*Arisaema quinatum*) has five-parted leaves, and I have seen it growing in a nearby valley.

Among the flowerless plants I recognize with certainty are wild

Jack-in-the-pulpits thrive in damp, shady places. There are several kinds. This one has three-parted leaves.

Less common than the previous jack-in-the-pulpit is the five-leaved species (Arisaema quinatum)) found in a few places in the mountains.

geraniums; I see a few of their leaves along the foot of the mountain. They rise above the earth on short petioles, their shapes re-
minding me of large, green snowflakes. Above this point the brook
is transformed into a series of miniature waterfalls tumbling down
from higher elevations. I now have the choice of following the
precipitous course of the stream or continuing my ramblings on up
the valley; I elect the latter, and am forced to scramble over large
boulders obstructing the way.

There is something of a sameness to the luxuriant vegetation of
the valley but, now and again, hidden away among the boulders,
there are often surprises. Here beside a fallen log, now completely
clothed with a mantle of green moss, I discern a cluster of pale
purple flowers of tubelike form. The blooms are located at the tops
of brownish stems completely devoid of leaves. These are cancerroots, one of the more unusual plants of the valley. It lives as a
parasite upon the roots of various plants.

The processes of evolution do, indeed, create strange habits in
plants, some that seem to defy explanation. Why, I wonder, would
a plant dispense with its green chlorophyll, when it could so easily
manufacture its own food? Still, here is a biological niche, one that
has been taken advantage of by a number of plants.

With my hunting knife I slice a circle in the soil around the
clump of cancerroots and lift them out. Then, walking over to the
river, I carefully wash away the adhering soil so as to expose their
root systems. The parasitic plants' attachments to the roots of other
plants are exposed. One cancerroot, I notice, is attached to the roots
of a violet, while others are attached to the roots of other, un-
identified, plants. Aside from a small cluster or rootlets, probably
serving to absorb moisture, the cancerroots appear to have no other
means of sustenance.

Far up the valley the road passes through open glades, then
squeezes between the river and the foot of the mountain, bending
around massive outcroppings of ancient stone. Some of the stone is
blanketed with green moss. At this season water from winter's
melted snow and spring showers percolates down through the soil,

Fringed phacelia grows in several places along the trail. Sometimes the blooms are tinged with lavender.

Droplets of morning dew glisten on the leaves of squirrel corn plants. The white blooms are seen near the middle of the patch.

dripping over the face of the stone, the falling drops sparkling in the sun and splashing upon the plant growth below.

Late in the day, I walk up the unused roadway, passing through glades almost as white as snow with fringed phacelia (*Phacelia fimbriata*). Some of the attractive little blooms are tinged with pale lavender. Over these flowery fields flit white butterflies, pausing now and then to sip nectar. Above the flower-strewn glades flutter occasional tiger swallowtails, hurrying on, apparently in search of larger blooms, disdainful of the modest phacelias.

Shortly, the road approaches the river that is now tumbling over a broad boulder-strewn bed. On my left the road bends around a rocky cliff at the foot of which is a rich growth of squirrel corn (*Dicentra canadensis*), its many-divided leaves still set with gleaming droplets of dew. Several of the squirrel corn plants are in bloom, their strange little flowers white against the pale green foliage. Farther up along the side of the cliff face I find a few Dutchman's-breeches (*Dicentra cucullaria*) with their little blooms

Closely related to squirrel corn are Dutchman's-breeches, their flowers resembling white pantaloons with yellow waistbands.

strung along on stems like upside-down pantaloons on clotheslines. Looking closely at the flowers of the squirrel corn and Dutchman's-breeches, it seems odd that both are closely related to poppies; they seem a far cry from the familiar poppies in my wife's garden. Yet I know from past experience that plant taxonomists often create strange bedfellows; it is, of course, the details of a flower's structure, not its gross appearance, that determines its true relationships.

Pulling up one of the little squirrel corn plants, I see the reason for its name; the rootstock bears numerous little spherical tubers of yellow color. These are assumed to be eaten by squirrels, and I presume that this is so. The early settlers, close observers of nature that they were, gave the plant its name and who am I to argue?

Beyond this first large, stony outcropping, the primitive road continues on along the river until it disappears some distance away around another bend. The day is warm and my camera case is heavy, yet I am unable to resist the lure of an unknown place. What, I wonder, lies around the next bend? As a boy in Montana I often explored a canyon high in the mountains. Alone, I would climb to a point where the dim trail rounded the foot of a steep cliff. Beyond lay an imposing mass of jumbled boulders, some as large as cottages. They represented the accumulated collections of stone tumbled down, during past earth tremors, from the sides of the steep mountain rising thousands of feet above. To me, it seemed an eerie, mysterious place and I often stood for long periods at the turn in the trail gazing at it in fascination. Yet I could never bring myself to move a step beyond that one point; it seemed as if an invisible wall barred the way. It was, perhaps, merely the fear of a remote, unknown place, yet I often explored other, even more isolated places in the Rockies without hesitation. Strangely, I feel now the same way about this bend in the road along Little River, but I know that I must see what lies beyond.

At the turn of the dim road the river continues on up the valley, and when my eyes sweep back toward the side of the mountains I am astonished to find it covered with a solid mass of blooming

A tiger swallowtail butterfly drinks nectar from a white trillium.

trilliums, each white, tripetaled bloom facing the morning sun. Not a single one faces in the opposite direction. Earlier in the morning down the valley other white trilliums were all facing in the opposite direction; that is, *away* from the sun. Here, I realize, is a small mystery, one that I am at a loss to explain.

Most of the trilliums on the mountainside have snow-white petals, yet occasionally there is one of pale pink hue. For awhile I am puzzled, then decide that these are probably older specimens and that some of the white trilliums turn pinkish just before fading. In autumn the fruit of this large-flowered trillium (*Trillium grandiflorum*) appears as a black berry nearly an inch in diameter. It is an interesting feature of the autumn woods.

Here, beside the road, I discover another trillium, this one a wake-robin (*Trillium erectum*). Its three petals are deep purple-red and the plant has a most unpleasant smell, which is the reason

The three petals of a wake-robin trillium are deep red. Because of its unpleasant odor, it is also called "stinking Willie."

A young cottontail hides in a clump of plants near the trail.

it is often known as "stinking Willie." Why it is called "wake-robin" I do not know; perhaps it is because of its habit of opening its bloom very early in the spring to "wake up" the robins.

The sun, now high in the sky, pours its warmth down into the valley in ever increasing intensity, causing steam to rise from boulders still damp from the cool night. Overhead a lone crow sails across from one mountain to the other, its form black against the sky. From the forest beyond the stream drift the songs of birds, their voices barely audible above the murmur of the rushing water. It is a day conducive to exploration and so I move on up the stream which, at this point, bends away to the right, leaving an area of dense woods between its channel and the road. This forested area of hemlock and other trees is damp and dark. It looks like a place where unusual plants might occur and so I scramble down the steep enbankment and push through the tall vegetation. In the half-light there are great logs, rotting and moss-covered, with vines trailing over them. A rabbit, flushed from its bed, darts away through a thick growth of ferns and, on a leaf, I see a leaf-cutter bee in the act of snipping out a circular disc. Having cut the disc from the leaf, the bee straddles it and flies away toward the mountainside where, I presume, it has its underground nest.

After the bee's departure I climb over a large log and unexpectedly come upon several beautiful specimens of yellow lady's-slipper orchids (*Cypripedium*). The great blooms rise nearly two feet above the ground, their large, expanded lower lips bright yellow against the green of the plants' broad leaves. These are the most attractive of the valley's wild orchids, always a source of pleasure to their discoverer. A short distance beyond the lady's-slippers I notice yet another kind of orchid, this time the showy orchid (*Orchis spectabilis*). There are several specimens, and I stop to examine them closely. They grow on much shorter stems than the previous ones, and there are several blooms on each one. The flowers have flat, snow-white lower lips, while the upper petals are purple and curve hoodlike over them. Most of the showy orchids I have previously found here in the valley grew on mountainsides;

Two yellow lady's-slipper orchids bloom in a damp place
near the river's edge. These
are among the most attractive
of the native orchids.

More common than the
lady's-slipper is the showy
orchid, with its purple upper
petals. Usually it grows on
hillsides.

Solomon's-seal grows near the bases of cliffs and in other damp places. A member of the lily family, its blooms are greenish-yellow.

thus I am surprised at finding these here on the valley floor.

In this same wooded area are numerous other flowers common in the valley. There is Solomon's-seal (*Polygonatum biflorum*) with its little bell-like blooms suspended in a row beneath the leaves. There is also false Solomon's-seal (*Smilacina racemosa*), often called "false spikenard," with its terminal cluster of delicate white blooms. Nearer the roadside grows a small patch of dwarf iris (*Iris cristata*), quite similar in appearance to the domesticated variety but with much shorter stems.

Among the rocks are clusters of delicate little bluets (*Houstonia*), and along the woodland's margin rise spikes of miterwort (*Mitella diphylla*), its tiny, cuplike flowers strung along verticle stems. Often known as "bishop's-caps" or "fairy cups," these attractive little flowers grow almost everywhere in the more humid parts of the valley.

Least esteemed of all the valley's blooms are those of the various grasses. Because of their small, inconspicuous size, few people appreciate the delicate beauty of grass blooms, yet when viewed under the magnification of a hand lens, their details are most attractive. Here I notice several grasses in bloom. Walking back down the road, I pause at several points to examine grasses of various other kinds, now in their flowering stages. Since grasses depend upon winds for the distribution of their pollen, they have no need for colorful or conspicuous petals to attract insects. Yet their numerous feathery pistils and yellow anthers, each anther trembling at the tip of a slender filament, are, truly, blooms in miniature and I marvel at their diminutive loveliness.

This brings to mind a fact of which I am always aware, emphasized by my habit of viewing so many of Nature's small wonders through the close-up lenses of my cameras. Most of us, I fear, are far too engrossed with large size; we marvel at the beauty of a lady's-slipper orchid yet ignore the dainty elegance of a miterwort, a grass bloom, or the details of a moss plant. The possession of a simple hand lens opens up a new and little-known realm, a lilliputian world well worth exploring.

Grasses of many kinds grow in the valley. The blooms are most attractive when viewed under a hand lens.

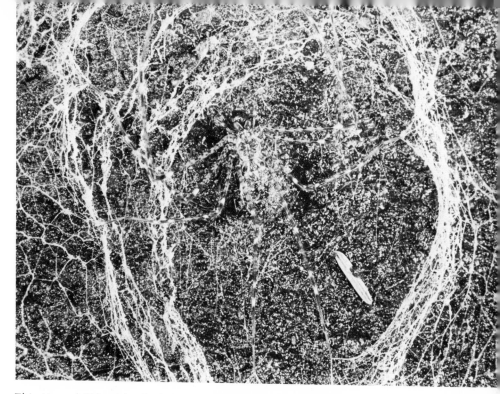

This Hypochilid spider built its circular web upon a boulder beside the river. Its mottled coloration makes it difficult to see against the lichen-covered stone.

On the way back down the river I leave the road to study a nearby wall of stone where the original strata are exposed to view. Upon the damp stone grow a number of ferns and other plants, their roots finding precarious anchorage in chinks in the weathered surface. While examining the plants I notice several spider webs attached to the rock and, resting in each one, a long-legged spider of mottled coloration. Each web is more or less circular in form, with an open center where its maker rests. I am astonished at the effectiveness of the concealing coloration of these spiders, their markings blending so well with the stone as to make them almost invisible. Later, I determined the spiders to be Hypochilids, a type found in several places in the United States, as well as in China and Tasmania. Usually their webs are built on the sides of boulders near streams, where they feed upon flies that become entangled in the silken webs.

Leaves of the basswood or linden tree show the tendency to form mosaics, leaving little unused space and not overlapping each other.

Chapter 4

LEAVES IN THE SUN

WHEN I AM AWAY from the Great Smoky Mountains the thing I remember most vividly about them is the profusion and variety of the leaves; they adorn every tree, every bush, vine, and shrub, as well as the lesser plants growing in their shade. Not even in tropical forests have I seen more lavish displays of green. The forest reaches up toward the sky in tiers, each living tree and plant striving upward as if drawn by some great magnetic force. This force, of course, is the sun, the source of the energy that enables them to live and grow, and come to fruition. Thus, it is not only local conditions that play a role in this abundant plant growth; the sun, 93,000,000 miles away, each day pours a steady stream of energy down upon the earth. In ancient times men justifiably worshiped the sun as the source of all life-giving heat and light. In a sense, most green plants are sun worshipers; they continually strive upward to take advantage of its energy-giving rays. It is an ancient struggle, as old as the first moss plant that crept across a boulder nearly half a billion years ago.

Everywhere in the forest there are evidences of the fierce competition for sunlight; the trees, stout of trunk and supported by great underground root systems, have won the battle, leaving smaller plants to do the best they can with what light filters down to them through the massed foliage of the trees.

As I walk through the valley forest I can almost see and feel the competition for light and energy. The leaves of the trees, in many

cases, are arranged in mosaics, each one so placed as to fill as nearly
as possible all the space, presenting to the sun an almost solid sheet
of green. Here, beside the path, is a small maple tree and when I
bend down and look upward through its branches I can see how
its leaves are arranged in such a way that each one tends to occupy
a definite space relative to the rest. This is brought about by the
leaf's placement on the twig and by the length of its leaf stem or
petiole. If I were to examine trees of other kinds I know that I
would find more examples of the way trees take full advantage of
available sunlight; some leaves spiral up the stems so that no leaf is
directly beneath the one just above it.

This competition for light is ruthless in the extreme, and it is
only through long adaptive evolution to life in dim light that the
lesser plants of the forest have been able to survive. On a clear day
at noon, the sun has a maximum intensity of more than 10,000 foot-
candles. Of this light, only about 30 per cent is within the range of
human vision, the remainder consisting mostly of invisible infrared
and a small amount of ultraviolet. Of the total amount of light that
falls upon a leaf, about 50 per cent is transformed into heat energy
used to vaporize the contained water, thus lowering the leaf's temperature; 20 per cent is reflected off the leaf and lost; only the remaining 30 per cent is actually used by the leaf in food manufacture. Inefficient as this seems, the trees and plants of the forest
continue to grow and to thrive. It also accounts for the fact that
the lesser trees and plants of the forest are able to survive in relatively dim light. Tall trees, whose leaves are exposed to the full
force of the sun, suffer considerable water loss. Their problem is
not light but excess heat. In shade-loving plants, on the other hand,
light is the vital factor and water loss through evaporation is
slight. As a result, there are internal differences between the leaves
of sun-loving and shade-loving plants. There are also differences
between the shaded leaves at the interior of a tree and the outer
leaves exposed to the full light of the sun. If a leaf from the interior
of a linden is compared with one from its crown, the latter is found
to be much thicker. In general, leaf thickness increases with light

Maple leaves are also so arranged on the twig as to form a mosaic, taking full advantage of available light.

intensity, but there are exceptions. For example, rhododendrons grow in many locations in the valley, often in very shady places. The same is true of dog-hobble (*Leucothoe fontanesiana*). Both of these shrubs have relatively thick, dark green leaves that reflect less light than light-colored leaves such as those of maple and poplar. These thick leaves absorb almost all the light that falls upon them, enabling the plants to survive in weaker light.

Each tree and plant in the valley forest is admirably adapted to the situation in which it lives. In rising above the earth toward the sun, the great forest trees have adjusted to conditions of excessive light, just as the shade-loving shrubs and herbaceous plants have become adapted to minimal amounts of light. Yet each plant and tree is also able to tolerate great variation in light intensity. In the forest's shade beside a large boulder thrives a goldenrod, now

The mottled limbs of the sycamore hold their leaves up to the sun, where they take advantage of the sun's energy.

in full flower. It is normally a sun-loving plant, flourishing in bright sunlight. Here, however, it receives hardly one-fifth of the sun's full strength, even at midday. So well is it adapted, however, that it can grow and bloom in much less light. Deeper in the shadows grow other plants—ferns, mosses, wild ginger, and violets—all of them thriving on as little as two or three foot-candles of light. This is about 1/3,000 as much light as is enjoyed by the tall forest trees.

Actually, the leaves of the beech produce several times more starch and sugar when located in the shade. On the other hand, it appears to make little difference to a hemlock whether or not it grows in direct sunlight, so well does it adjust to variations in light intensity. Many plants, of course, are especially fitted to life in deep shade and cannot grow where there is too much sun. Ginseng, a plant whose roots are believed by some to have medicinal value, grows in the valley and it is always found in deep shade. Sometimes it is grown commercially, but in such cases it must be protected from the sun by slat-frames.

Unconsciously, I suppose, I am always comparing these mountains with the Rockies where the vegetation is relatively scant in most places. These latter mountains, higher and far younger, are bathed in about the same amount of sunlight, but moisture there is the governing factor, coupled, of course, with lower temperatures. The Rockies receive perhaps twenty inches of precipitation a year as compared to more than seventy inches in Little River Valley. This is the secret of the abundance and wealth of the flora of the Smoky Mountains; they enjoy ample sunshine and large amounts of moisture. Temperatures, too, are higher than in the Rockies. All

This young stonecrop's leaves are so arranged that none is shaded by another. These plants usually grow on the tops of boulders in the forest's shade.

these factors are conducive to luxuriant plant growth.

Leaves, seemingly in infinite number, festoon the trees and herbaceous plants of the valley, and one afternoon I wondered how many there actually were. My first thought was that it would be impossible to make even a wild guess. Yet when I considered the matter, it occurred to me that by calculating the number of leaves per square foot—not an impossible task—I might arrive at some reasonable figure. On the other hand, the leaves seemed to be illimitable in number and, looking away through the forest, I was almost convinced that the task would be far too complicated.

In any case, I decided to attempt an estimate of the number of leaves—on both trees and herbaceous plants—on an acre of ground in Hidden Valley. I imagined a column, one square foot in area, reaching upward from the earth to the tops of the trees and estimated the number of leaves within it. A month later, after the leaves had fallen from the trees, I made several counts of dead leaves on the ground and obtained an average. The conclusion was that on each square foot there had been an average of about two hundred living leaves. The conifers—hemlocks and pines—I ignored, since I could not decide how to classify their needles.

From the above figure I determined that on each acre of ground there had been 8,712,000 leaves. Carrying my calculations even farther, assuming there to be about six square miles in the valley, I found that there had been about 33,454,080,000 leaves. It occurred to me that this is less than the number of dollars in the national debt. As a result, for some time after making these calculations I was conscious of the comparison between leaves and money. Each time I walked through the valley forest I saw not leaves but dollar bills suspended from every twig. I was fascinated in discovering just how vast is a billion, be it leaves or dollars. When I know that in a hundred acres of jungle-like Smoky Mountain forest there are approximately a billion leaves of all types, I begin to appreciate the enormity of the number. Written figures mean little to the human mind, but leaves are concrete entities that can be seen and touched. And so, each day, as I wander through the forest, seeing the myriads of individual leaves, I understand perhaps a little better just how large a billion really is. The word "myriad" comes from the Greek word *myrios*, meaning "countless," and yet the leaves of the valley forest are not countless; in a manner of speaking, I have counted them.

Record sizes of things are always of interest, whether they be for bigness or for smallness. The world's largest leaf is that of the royal water lily found on the Amazon River. Its leaves are round in shape and measure up to twenty-one feet in circumference, or about six and one-half feet across. Here in Hidden Valley the largest leaf—admittedly small by comparison with the royal water lily —is that of the Fraser magnolia, which often reaches a length of nearly two feet. Which local plant has the smallest leaf is difficult to say. Floating duckweeds (*Lemna*), found in ponds, with minute leaves less than one-eighth of an inch across, are the smallest of all flowering plant leaves. However, there are no ponds along Little

Probably the largest tree leaves in Hidden Valley are those of Fraser magnolia, which are often nearly two feet long. The umbrella magnolia, with even larger leaves, grows at lower elevations.

The largest leaves of any herbaceous plant in the valley are probably those of the umbrella leaf (Diphylleia cymosa), which grows in damp situations.

River and so I doubt that duckweeds occur here. I cannot even guess which local plant has the smallest leaves.

Abundant as are the leaves of these forests, each one has its own individual form and structure; no two, even on the same tree or plant, are exactly alike. Between the different species the variation is, of course, even greater. Those of the rhododendron, as we have noted, are thick and leathery. Others, such as those of the linden, are very thin. In my library at home I have a publication by a botanist in Brazil, Dr. Carlos Toledo Rizzini, in which leaf shapes are classified into fifty-five different categories, ranging from orbicular (round) to stellate (star-shaped), cordate (heart-shaped), and filiform (grasslike). Leaves are also simple and compound. Almost every shape that the human mind can conceive is represented in the classification and I am sure that if I had the time I could find examples of every type here in Hidden Valley. The leaves of the ferns alone would include many examples and illustrate that even in a single plant group there are leaves of amazing diversity. Someone has said that God created ferns to show what He could do with leaves.

Why, you may ask, are there such variations in leaf shapes? The answer is not at all simple. Some leaves have pointed, downwardly directed tips that facilitate the runoff of rain water, eliminating the water before it can injure the leaves by inducing the growth of fungi or by focusing the rays of sunlight upon the leaves' delicate tissues. In the forest there are many examples of leaves with drip tips. On the other hand, many leaves are ovate in form, having no adaptations for the rapid elimination of water. Such leaves, as nearly as I can determine, function just as well as those so equipped. The large, cordate leaves of wild ginger often rest upon the damp ground and have no provision for water runoff. Yet these plants flourish, suffering no visible injury. The leaves of the scarlet oaks are equipped with several drip points, whereas white oak leaves have rounded lobes with no points. Yet one tree seems to thrive as well as the other.

The subject of drip tips is an illusive one and I hesitate to generalize too much. Drip tips must have value; otherwise not so many

The leaves of ferns are of almost endless variety. This tapering fern is common in the valley.

leaves would be equipped with such a mechanism. I recall that the buckskin jackets of the American Indians and early trappers were almost always fringed. These fringes, contrary to the usual assumption, were not merely decorative; in effect they were drip tips, aiding water to drip off quickly without soaking the remainder of the clothing.

During my peregrinations up and down the valley I often ponder this question of variation in leaf shapes, but there seem to be no logical reasons for most of them, any more than we can rationalize the coloration of most birds; they are merely another facet of the complex operation of evolutionary forces tending to differentiate one plant or animal from another. I once engaged in an interesting conversation with a ship's doctor while cruising through tropical seas. Seated in the ship's wardroom after the evening meal, our discussion became, in some way, involved with Nature's beauties. "Why," the doctor asked, "are most plants and flowers attractive to our eyes?"

I could not answer his question; my only conclusion, after some thought, was that Nature tends toward the symmetrical and the beautiful. When thought of the other way around, I wonder if a flower or a leaf is attractive to our eyes because we, long ago, became conditioned to regard them so? They were already present when we emerged from savagery; perhaps we looked at them and

The leaves of wild geranium look like large, green snowflakes on stems.

found them attractive because they were the most colorful and symmetrical objects visible in the natural world around us.

Many trees and plants of the valley, especially the herbaceous types, show another interesting characteristic. The lower, older, leaves differ from the topmost or younger leaves in that they have more divisions. (Queen Anne's lace is an example.) Botanists, however, have found that it is not the age of the leaves that accounts for the variations in shape, but their positions on the stem. Another phenomenon can be seen in the leaves of very young trees; their leaves, in many instances, are larger and of a slightly different shape from those on mature trees.

There is also the matter of air flow over the leaf. A leaf cut into many divisions, it is believed, functions more efficiently than one of solid form. According to this theory, air flows more readily over a divided leaf such as that of Queen Anne's lace, with the result that its respiratory processes operate more efficiently. This may account for the cut or divided form of many leaves. Yet we can easily think of a hundred examples of trees and plants with undivided leaves that apparently get along just as well.

According to the views of some plant physiologists, shade leaves are more apt to be undivided than sun leaves, and in examining the leaves from a number of shade-dwelling plants in the valley I find many examples in support of this theory.

The surface of an average-sized leaf contains about four million cells. Deeper within the leaf there are numerous cells of other kinds. To the unaided eye the leaf appears to be solid green, with many raised veins running across it, yet if we look at the leaf's cells under high magnification we find that their green coloration is not uniformly distributed. Instead, it is contained within numerous microscopic bodies known as chloroplasts, each one only about 2/10,000 of an inch in diameter. The green pigment in these bodies is commonly known as chlorophyll, but there are, in truth, several kinds. The most abundant one is chlorophyll a, but other forms—b, c, d, and e—may also be present. Always associated with the chlorophylls in a leaf are other pigments called carotenoids. One of these

The leaves of agrimony are unusual; small leaflets appear between the larger ones. Is this an adaptation to help them take advantage of available light?

is beta-carotene, another is xanthophyll. Beta-carotene, the most abundant, is orange-yellow and when eaten by animals is changed into vitamin A. Carotenoid pigments are responsible for the autumn coloration of leaves as well as for the bright hues of many flowers, fruits, and seeds. Fortunate, indeed, are we who live near, or can visit, the great deciduous forests of North America when they display their extravaganzas of color at summer's end. People living in many other parts of the world are not so privileged. Autumn coloration is never seen in Africa, and in South America only in southern Chile. In general, this yearly display is rare south of the equator. It does occur in the British Isles, Europe, and parts of Japan, as well as in our own country.

Here in the Great Smoky Mountains autumn colors reach, perhaps, their most extravagant hues, due to the fact that this is the supreme habitat of eastern deciduous forests. October is the time when these mountains take on the brilliant panoramas of color so thrilling to the many visitors to the area. Cool nights and bright days at this season bring about the physiological changes in the leaves that result in the production of the gay, vivacious colors of autumn.

At first, the oaks remain green, as if reluctant to don their autumn garments. In the meantime, the sourwoods, dogwoods, and sumacs turn scarlet and crimson. The poplars and hickories turn golden-yellow. Persimmons are dressed in royal purple. Gradually, as the season progresses, the coloration of the forest changes; the scarlet oaks, at last, attire themselves in cloaks of flaming crimson, and the mountainsides glow with living color.

All these colors result from the presence of the pigments previously mentioned. These pigments comprise only about 3 per cent of a leaf's weight, yet impart to it all its color—the green of summer and the yellow, scarlet, or purple of autumn. The carotenes are responsible for such leaf colors as yellowish-orange and red, as well as for the colors of sunflowers and goldenrods. Xanthophylls are more abundant in tree leaves than the carotenes and impart

Each leaf of the forest is an efficient manufacturing center where sugars and starches are produced. The leaves of deciduous trees like this maple are discarded in autumn and new ones unfold in spring.

yellow colors to them. Thus, yellow autumn coloration is usually dominant.

The two carotenoid pigments—carotene and xanthophyll—are present in the leaf all summer, but at the approach of autumn a new class of pigments appears. These are the anthocyanins which impart to the leaves in which they occur various brilliant hues of scarlet, purple or, even, deep blue. The colors of many flowers are due to these same chameleon-like pigments. Chicory flowers, when freshly opened, are light blue but later change to pale pink. These pigments also impart to roses their deep red and to cardinal flowers their scarlet. They also give blue violets their coloration, as well as the hues of grapes and blueberries. They are also responsible for the attractive red color of ripe apples.

Another type of pigment found in many leaves, as well as in walnut hulls and in the bark of many trees, is tannin (sometimes called tannic acid). This substance is very bitter to the taste and is responsible for the bitterness of unripe persimmons. The brown or golden-bronze colors of some oaks are caused by the presence of tannin.

Most people credit Jack Frost with the gay colors of autumn, but this is largely in error; there are many other factors that have more influence. Pigments are affected by light intensity, air and soil temperatures, moisture supply, as well as other things. Light enables a leaf to produce more sugar which, in turn, affects pigment formation. Temperature is of paramount importance, but the most favorable range for color production is slightly above freezing. Freezing temperatures, which trap sugars and tannins in the leaves, simply kill the leaves before color is produced. Thus, cool but not freezing nights are most favorable. As a general rule, most bright autumn colors are especially evident during years that are sunny and dry, followed by rain in early fall and, later, by cool, but not freezing nights. Cloudy, warm autumns usually result in dull colors, since sugar production is reduced. On the other hand, yellows and browns are little affected by autumn weather.

The usual autumn colors and the pigments that cause them are given below for the common trees of Hidden Valley:

	AUTUMN COLOR	DOMINANT PIGMENTS
Beech	Pale gold	Anthocyanins and tannin
Birch	Golden-yellow	Carotenoids
Dogwood	Crimson or scarlet	Anthocyanins
Elm	Butter-yellow	Xanthophyll
Hickory	Dull gold or yellow	Tannin and xanthophyll
Locust	Yellow	Xanthophyll
Maple	Scarlet	Anthocyanins and carotenoids
Sassafras	Yellow, pink, or blood-red	Anthocyanins
Scarlet oak	Red	Anthocyanins and tannin
Sourwood	Scarlet	Anthocyanins
Sweet gum	Crimson, yellow, or purple	Anthocyanins
Sumac	Red	Anthocyanins and tannin
Sycamore	Yellow or golden	Xanthophyll and anthocyanins
Tulip poplar	Bright yellow	Xanthophyll
White oak	Purple-red to violet	Anthocyanins and tannin

Autumn in the valley is probably my favorite time of the year. With due respect to spring, when the floral display is at its best, I still prefer the end of summer, the period before winter actually settles down in the mountains. Days in Hidden Valley at this time are clear, brilliant, and bracing, and the nights are cold; often the thermometer drops down to freezing and at dawn the landscape is white with frost. On such mornings, however, I have often been faced with a photographic problem; forgetful of the cold nights, I have frequently left my cameras in my car, with the result that the lenses fogged upon exposure to the damp, morning air. The only solution has been to place the cameras in front of a heater until their lenses were warm. I have also had trouble with electronic flash rigs in the cold and dampness. Temperamental in the extreme, these lighting units often fail to flash at crucial times and may require several hours' exposure to the hot sun before they will function as usual.

During autumn days the chipmunks, always alert and vivacious, are even more active than usual, hurrying about on the ground among the sun-warmed leaves, gathering acorns, buckeyes, and other seeds. They scurry away with their prizes, darting into underground dens.

Spring is the season of beginnings, when flowers come into bloom and the trees of the valley forest unfold their leaves. Spring is the season of youth, filled with the promise of the future. Autumn, by contrast, is the time of fruition, the climax of all the growth forces of Nature that have been active all spring and summer. Early blooms have long since gone, while on exposed hillsides asters and goldenrods now come into flower, pouring their life-forces into the production of blooms and seeds in preparation for the continuation of the species during another year. It is an ancient cycle.

Seated comfortably on a great boulder beside the river, I see the last black swallowtails of summer fluttering down the stream. Now and again they alight upon purple asters along the bank, sipping nectar, then flying on again. They are fugitives from summer; their usefulness ended when they laid their eggs on Queen Anne's lace, thus assuring that their offspring will emerge from chrysalids when the warmth of another spring comes to the valley.

In the surrounding trees are migrant birds, colorful warblers of many kinds searching each leaf and twig for insect food. But they are birds of passage; they will remain for only a brief time, then move on toward more southern climates. Floating slowly over the river in the golden light of autumn are the silvery strands of spider silk. They are wafted gracefully above the water, undulating in the gentle breezes blowing down the valley. These silken strands are silvery balloons spun by tiny spiderlings to carry them lightly through the air. The spider clan never evolved wings, yet they sail through the air along with the air-borne seeds of dandelions and milkweeds.

More eye-catching than anything else are the falling leaves. Golden and scarlet, some of them land upon the water, gliding across the dark pools and tumbling over the small cataracts, then

During early spring and late autumn, most of the trees are bare, allowing the sun to bathe the forest floor, stimulating the growth and blooming of the lesser plants growing there.

During the late spring and summer, the trees are clothed with foliage, shading the plants on the ground at a time when the sun's rays are most powerful.

In autumn, leaves often fall into the river, floating lightly upon the surface. Here, they appear to grow upon a tree mirrored in the surface of a pool.

floating on again. In time they will become waterlogged, their tissues decayed. They will add their organic matter to the water of the great river into which this stream eventually flows. Each airy gust brings down more showers of colorful leaves that splash lightly upon the water and coast past my vantage point. One by one, they are being discarded by the trees they served so well during the time of active growth. Now, like old, worn-out articles of raiment, they have been cast aside, useful to their owners no longer. And yet I watch them fall and drift away with a feeling of nostalgia; I remember when they unfolded in the spring, each one a bright new blade undamaged by wind or insect. During the summer days they were busy manufacturing starches and sugars, and piping these foods away to the twigs, limbs, and roots. Now, at summer's end, the nutrients begin to drain away from the leaves, to be stored in the branches. The leaves are now filled with accumulated waste materials and, being soft and tender, cannot survive the cold of winter. Ice crystals would rupture and destroy their cells and so render them unfit for use. And so, by an orderly process, certain cells at the base of each one form a cutoff zone called the *abscis-*

sion layer. Soon, the walls of the cells composing this zone begin to disintegrate and the leaf eventually falls in a gust of wind.

Trees that discard their leaves in autumn are said to be deciduous, but not all deciduous trees lose all their leaves at once. The leaves of red oaks remain on the trees throughout the winter, probably because an abscission zone is not formed or is not complete. Many other trees and shrubs of the valley do not discard their leaves at summer's end. The needle-like leaves of the pines and hemlocks are filled with resins and thus can survive the cold. The leathery leaves of the rhododendron roll up when the temperature falls. With exposed leaf area thus reduced to a minimum, they will not be injured. In spring they will continue to manufacture food.

Seated here on a boulder this mid-October afternoon, I watch the falling leaves sailing down like gayly colored confetti and marvel at the miraculous autumn season and the indiscriminate array of colors around me. All the mountainsides are ablaze with multi-hued glory. The trees—birches, sourwoods, and maples—hang over the stream like a colorful canopy, bathed from above in the golden light. Each gust of air brings down more leaves, and as I watch I see each kind sailing down in a characteristic manner. Usually I can identify a leaf by the way it falls, although the shape in which

By autumn's end, most of the leaves of the forest have fallen to the ground, discarded by the trees they served during summer.

it dries before falling from the tree also influences the path it follows. In general, maple leaves spiral downward, following a helical path; oak leaves zigzag in their descent, swinging from side to side in hurried movements; the leaves of the sycamores settle gracefully down, exhibiting but little lateral movement and do not spin. (Sycamore leaves remind me of small, inverted parachutes.) Willow leaves, slender and lanceolate in form, have a most characteristic manner of fall; they spin rapidly on horizontal axes. I am sure I could classify each kind of tree leaf by the way it falls. Each one, by its shape, is governed by the complexities of its aerodynamics.

Against the background sounds of the roaring stream in Hidden Valley is the music of the forest, the multitudinous voices of the trees as the winds blow through them. There is the soft but audible breath of the breeze in the pines and the hemlocks, and the sonorous tones of the broad-leafed trees. Never is there complete silence in the valley, and often, while alone there, I imagine that each tree has its own special "voice." Certainly each one sounds differently when the wind blows through it. I feel sure that I could tell in which part of the world I was, merely by the sounds of the local forest. In the tropics the palms emit a special tone, the soft hiss of the trade winds blowing their waving fronds. Often, while walking quietly through the jungle-like hammocks of the Everglades, I have listened entranced to the characteristic rattle of palmetto

The cinnamon fern usually grows along streams and in other wet places. Its spore-bearing frond, center, is brown.

fronds rasping harshly one against another. No other forest that I know has such sounds. Again, on the slopes of high, wind-swept western mountains, I have thrilled to the hiss of winds in the piñon pines—sounds that were somehow modified and exaggerated by distance and elevation. But always in Little River Valley I seem acutely aware of the voices of the trees and of the tumbling stream, their special quality and tone. Even with eyes tightly closed, I know that I am there.

The human mind is an attic wherein are stored the accumulated impressions of a lifetime. There, hidden away, our memories gather dust, but now and then we may withdraw them, letting them slip through our fingers like a miser gloating over his gold. Most memories are pleasant and so I bow in reverence to Mnemosyne, the goddess of memory in ancient Greece.

When away from these mountains, I see in my mind's eye a delicate fern growing in a crevice beside a woodland waterfall, droplets of spray sparkling in the morning sun with prismatic color. I recall a scarlet maple high on a mountainside, surrounded by the dark green of the conifers and glowing like fire in the autumn light. I see a lone sourwood, its foliage deep scarlet and its branches festooned with rows of seed pods. I remember, too, looking upward through the spreading branches of another scarlet oak, its leaves glowing fire-red against the October sky. Sometimes I see bright leaves of autumn floating slowly across a dark woodland pool in the fading light of evening.

Above all, I remember the multitudinous leaves of the valley, their slow unfolding in spring, when the snowy dogwoods are in bloom and the leaves tender and as yet undamaged. I see again the massed verdure of the mountains during the hot days of summer and hear the sound of the wind in the pines. With special pleasure, I live again through the bright days of autumn when the mountainsides suddenly burgeon with riotous color, and when the earth is blanketed with its Jacob's robe of varicolored leaves. These are among my store of recollections of days along the river. Time can neither dim nor erase them.

The waters of Huskey Branch rush down from the mountainside between boulders covered with mosses and lichens. On either side stretch dense jungles of rhododendrons and other shrubs.

Chapter 5

THE WATERFALL

IN MY RAMBLINGS AFIELD I seem always to be seeking isolated spots, hideaways visited by few, where I feel remote from the world and where I may ponder, perhaps, on life's meaning and, hopefully, catch glimpses of wild things going about their normal ways. On a South Pacific island, I found such a place; it was hidden high above the sea on the side of a mountain and enclosed by a stand of giant bamboos, their great, grasslike stems forming walls on three sides. At a later date, upon "my island" hidden in the vast Pascagoula swamp along the Gulf of Mexico, I found another hideaway, this time among the great, spreading branches of an ancient live oak.

Here, in Hidden Valley, there is a place where a small but most attractive waterfall drops down the mountainside and passes, at last, beneath a bridge that spans its course. The foaming water rushes down from ledge to ledge, splashing upon the moss-covered rocks where the droplets gleam brightly for a time, until absorbed by the living carpet of mosses and ferns.

This stream is known as Huskey Branch, and at the point where it rushes into Little River there is a deep, quiet pool, its rocky bottom clearly defined through the pellucid water. Only at midday does the sun bathe the pool in its direct light, but always it is the habitat of trout. In spring the inverted images of blooming dogwoods are mirrored in the surface along the opposite shore, while in autumn the forms of leaves, golden and scarlet, float gracefully along in the slow-moving current or congregate along the pool's sandy margins. It is place of quiet beauty.

On my frequent trips up the valley I have often paused to drink in the sense of peace and quietude prevailing at this spot, thankful that so few people have found it. In early spring a number of visitors drive up the valley to see the wild flowers and in autumn they come again to marvel at the autumn colors. During the remainder of the year the upper valley is almost deserted except for a few fishermen.

Several times I have wondered where the water comes from before tumbling down into the river. Looking upward, I could see that on either side of the falls there were steep declivities covered with jumbled boulders, but beyond this lay an impenetrable tangle of dense vegetation—rhododendron, birch, great oaks, and other forest trees, all towering up toward the sky.

This morning as I walk up the valley, the time seems proper and I am in the mood for exploration, so I leave the road and climb upward over the mossy boulders along the left-hand side of the waterfalls. At first, the way is fairly easy; I am able to scramble over the declivities with but little effort by bracing myself against the stony surface while grasping the branches of overhanging rhododendrons. Unfortunately, the higher I climb the more difficult becomes the way. Fifty feet, perhaps, above me is the first ledge over which the water spills before its next fall. For a minute or so I rest to regain my breath. A catbird meows from beneath a bush, its call muted by the continuous murmur of the cascading water. All else is quiet and there is no other sign of life.

Lured on by the unknown, I continue my climb, finding the going more and more difficult, but determined to see at firsthand what lies beyond view above the waterfall. It is a mystery of a sort that needs exploring and so I scramble on upward, over and around the great boulders. Alongside my precarious way the water spills down from ledge to ledge, while between each fall there is a quiet pool. Looking into one of these pools, I discover a number of caddis insects dragging their pebble cases along over the sandy bottom.

Beyond the quiet stretch of water, Huskey Branch drops over a ledge, its water spotlighted by a beam of sunlight.

One kind has built its cases of small twigs; other cases are con-structed of sand grains and are square in cross section. On the water's surface several water striders float quietly, their shadows projected upon the bottom. Here, no doubt, these insects are safe from fish; by no stretch of my imagination can I picture a fish ar-riving at this isolated pool far above the river. Yet, even here, there are the hunters and the hunted; the water striders capture and eat tiny gnats and other small insects. Perhaps they are even able to capture the long-legged crane flies that dance in shady places near the water. Rolling over a flat stone in the pool, I see a number of stone fly nymphs; they, too, are hunters, living by feeding upon small items of aquatic life. And so I know that in the quiet world about me there are little unseen dramas; pulsating life is going on, quietly moving forward with the cycle of the seasons. The time is autumn, and winter will soon arrive at this remote spot, bringing

Water striders skate over the quiet pools on slender legs, making small dimples in the surface film which appear as shadows upon the sandy bottom. Below: Upon surrounding leaves perch stone flies, emerged from nymphs that lived in the pools and fed upon small items of aquatic life.

In one pool are caddis insects that have built cases by cementing small twigs together. In time, they will transform into mothlike caddis flies. Below: Long-legged crane flies dance over the dark pools, alighting sometimes on the surrounding vegetation, this one on a hemlock.

its cold and snow. The caddises will withdraw into their stone fortresses and the water striders and the dancing crane flies will long since have laid their eggs and perished. The falls will be encased in ice, beneath which the water will continue to cascade down, its sound muted.

While I quietly contemplate the scene, a chipmunk scurries out from behind a nearby boulder, looks in my direction a moment, then picks up a seed and hurries away with its treasure. Somewhere nearby, I am sure, it has its underground den. As the autumn days grow shorter it will continue to stock its storehouse with food—seed and nuts—against the time of need. After the chipmunk disappears I continue my climb and am able at last to peer over the brink of the topmost fall. Surprisingly, the stream beyond this point levels off and, before dropping down over a distant ledge, it flows for some distance over a level bed between high banks bordered by impenetrable growths of rank vegetation. The distant ledge is hardly more than two feet high, but the water foams over the brink and drops into a pool with splashing sounds. Beyond this, the stream is hidden from view and I can only guess that it continues in more or less the same direction back into the mountains.

Shielded by a large boulder, I survey the attractive scene, my eyes gradually becoming adjusted to the semigloom. As I pause here in this isolated spot, I can hear the murmur of the waterfall and the distant call of a bird in the valley below. But there is another sound, and at first I cannot determine its origin. Slowly I become aware of twittering noises and then see a group of small birds, perhaps a dozen, clustered on the sandy margin of the stream. They are about fifty feet away, but as nearly as I can determine, they are chipping sparrows. They seem to be very much excited about something, but what it is I cannot at first see.

Cautiously I continue to peer over the boulder, watching the little tableau in the glade. The birds twitter and flutter their wings in obvious excitement, totally unaware of my presence. I am, indeed, mystified by the actions of the little birds, so I wait quietly, hoping to determine the cause of their excitement. Suddenly, they

At the brink of one waterfall is a stone covered with green moss where a lone violet blooms.

all hop sidewise, revealing the reason for the turmoil; it is a black snake resting quietly upon the sand, facing the birds and only a foot or so beyond them. As I watch, the birds, perched together in a compact group, hop toward the snake, then hop backward, repeating the performance again and again. I am reminded of a troupe of dancers going through a routine on a stage. The birds are highly excited, but the snake lies immobile on the sand, seemingly oblivious to their presence.

Fascinated by this little spectacle, I continue to watch, keenly interested as to how it will end. The snake remains aloof, almost disdainful, it would seem, of its tormentors. It is not coiled and thus probably cannot strike at them; it rests upon the sand almost as if dead.

For nearly ten minutes more the sparrows continue their hopping and twittering. Sometimes they all advance toward the snake, then retreat, always facing it but never nearer than about a foot. They seem to know that the snake cannot strike at them at this distance.

At this point a blue jay alights upon the branch of a rhododendron above the stream and adds its voice to the tumult. I am not at all surprised at this intrusion, since a jay's very nature is to become involved in every act. Understandably, the snake is now disturbed by the general turmoil; it begins slowly moving across the sun-dappled sand toward a fringing growth of ferns. At first, the birds retreat as the snake advances, then they all take to the air and alight in nearby bushes. In the meantime, the jay has become even more vociferous; it hops from limb to limb in excitement as the snake disappears in the lush vegetation beyond the stream. The sparrows, however, have not lost interest; they hop from one bush to another above the snake as it crawls along the ground. The snake is now hidden from my view, but I can easily follow its progress by the sounds of its tormenters. Apparently the snake is crawling up the mountainside, and the twitterings of the birds fade away in the disance. At last, I am left alone in the glade, having been an eyewitness to a most amusing little woodland drama.

I wonder why the sparrows were so strongly attracted by the snake. Snakes are birds' enemies, often raiding their nests and devouring their eggs or young. Were they "charmed" by the snake, or did the mere presence of danger hold some fascination for them? Perhaps bird psychologists may know the answer; I do not.

Now that the snake and the birds have gone, I suddenly realize that, in my hand, I hold a camera with which I could probably have recorded the entire episode. I had been so engrossed by what I had seen that I had completely forgotten it. There is the consolation, however, that the first click of the shutter would no doubt have frightened the birds away and so I would not have been privileged to see the entire affair.

Climbing over the moss-covered ledge behind which I have been hiding, I walk on along beside the stream, noting the little cluster of bird footprints and the snake's path in the damp sand.

On later trips to this hidden spot I discovered that for only a short period of each day does the sun shine directly down upon the small woodland stream at this point. During other times the conformation of the mountain keeps it in shade. Today I am in luck;

the sun now just touches the stream and the waterfall beyond the
quiet stretch, back-lighting the tumbling water, the splashing droplets sparkling brightly against the dark stone and upon the ferns
growing there. Along the stream the slanting light falls upon spider-
webs glowing silvery against the deep forest shadows. At the center of each intricate web rests its creator, immobile, waiting for
prey as is the way of spiders. There are several kinds, the most
abundant of which are the microthenas (*Microthena sagittata*)
with their spiny, arrow-shaped abdomens. A short distance away,
resting quietly on a leaf, is a lynx spider (*Peucetia viridans*). Bright
green in color, its long legs set with black spines, it is feeding on a
small beetle. Unlike its relatives, the orb-weaving spiders, it does
not depend upon a silken snare to obtain its food. Instead, it waits
patiently until some unwary insect strays too near, then leaps out
and makes its capture.

Close beside the water there grows a large jack-in-the-pulpit, its
odd-shaped flower green against the dark moss carpet along the
bank. Standing upright beneath the overhanging hood of the flow-

*A lynx spider, vivid green with black dots, feeds upon a ladybird beetle
near the waterfall.*

er's vaselike spathe is the white spadix, a structure that gave the plant its imaginative name. These aroids are common everywhere in the Smoky Mountains and so the finding of one is not unusual. Yet I am always intrigued by these plants. While observing the jack-in-the-pulpit, my eyes are attracted by what appears to be an ant crawling slowly across one of its leaves. Ants of one kind or another are common everywhere, and so my interest is not particularly stimulated. Yet, there seems to be something a little out of the ordinary about this particular specimen and so I look more closely, astonished to find that it is an ant-spider (*Synemosyna*), perhaps one of the most remarkable of all members of the great spider tribe. Antlike in body form and coloration, the little creature hurries about over the leaf, its actions quite similar to those of an ant. Unlike spiders, it seems to have six legs instead of the usual eight. In evolving its ant mimicry, these peculiar spiders were faced with a problem; presumably, its enemies or its prey could count up to eight and so the possession of eight legs was a dead giveaway, destroying the illusion that it was an ant. The problem was what to do about the extra pair of legs. But Nature, always resourceful, was able to solve the problem in a most satisfactory manner. The spider, in time, evolved the habit of holding its two front legs out in front of its body like antennae, leaving six walking legs, and so the illusion was complete. At first glance, these spiders resemble ants very closely and their movements are also antlike; they never jump, do not spin webs. Instead, they crawl about as is the habit of most ants. What, I wonder, is the advantage of this elaborate disguise? Does it aid the creature in capturing prey or does it afford protection from enemies? To a hungry bird, is an ant less desirable than a spider? Certainly there must be some advantage, otherwise evolutionary processes would not have created such a creature. I do not know the answers to the questions and so I leave the antlike spider crawling about over the jack-in-the-pulpit leaf and walk on up the stream, climbing over the little waterfall and pushing on between the bordering walls of plant life.

For some distance beyond the waterfall the vegetation reaches

On a jack-in-the-pulpit leaf crawls an ant-spider, a spider that mimics an ant. It holds its front legs above its head like antennae.

completely across the stream from either side, shielding it from the sky. Progress is difficult but I move on up the channel, stepping from stone to stone, each topped by a green clump of moss. As I go onward, the sounds of the water dropping down to the river fade away behind me and I find myself in a green, leafy world surrounded by great forest trees all laced together by pendant grapevines coiling up toward the sun, their tendrils groping for anchorage on every available twig. Their thick stems twine about the trunks of the trees, often reaching across from one to another in graceful loops. Climbing up the trees, too, are vines of the Dutchman's-pipe, their ovate leaves green against the dark bark. If it were spring, their strange, pipe-shaped blooms would be suspended along the stems. But the blooms have long since gone, leaving only a few seed pods. After the first frost of autumn the thin leaves, too, will be gone, fallen to the earth as the forest takes on its winter aspect.

Perhaps a hundred feet up the stream I find that its bed widens again and once more I can walk along on an open, sandy shore between the water and great moss-covered boulders. I am now completely isolated in the mountain forest, my view restricted to

the nearby bordering vegetation rising like walls around me. I pause here a moment, pleased at finding this sylvan retreat and enjoying the delusion that I am the first man ever to stand in this spot. Far above I hear the sound of a bird and shortly a tufted titmouse drops down and perches on a swaying vine. It eyes me for a few seconds, then flies away, the only movement in the vicinity except for water skaters skipping about over the slowly flowing water. And so I walk quietly on, noticing a fresh heap of bear droppings on the sand beneath a rhododendron. This, of course, is a little disturbing; I have no wish to encounter a bear in this spot where retreat would be difficult. I must confess that I do not like meeting bears in the deep forest, even though most of them are not supposed to be dangerous if let alone. Certainly I have great respect for these animals and have no intention of irritating one in its own domain. I search the vicinity and listen for sounds of movement but neither see nor hear anything, so I walk on up the stream which curves toward the right and disappears from view perhaps a hundred feet away. At the point where it vanishes there is a large boulder, its top covered by a dense growth of stonecrop. My curiosity still active, I continue along the stream, my feet treading on the damp sand. I reach the large boulder very quietly and hesitate before going around it. At this point I hear low shuffling sounds coming, apparently, from beyond view around the bend in the stream. Still bear-conscious, I stop, undecided whether to continue on or retrace my steps. However, having come this far, it seems foolish to turn around, so I bolster up my courage and approach the boulder, having no idea what to expect, but fearing the worst in the form of a large black bear. At first I see nothing, then near the left bank, beneath a tangle of doghobble vines, I discern a large skunk busily engaged in digging in the soft earth. It seems completely oblivious to my presence as I watch quietly from my vantage point. It is a beautiful specimen, the upper part of its body almost completely snow-white, and I can even see its black beadlike eyes each time it withdraws its head from the hole it is digging. Completely hidden as I am behind the screening ferns, I feel quite certain that it is not apt to see me, so I continue to watch, unable to guess what the creature is up to. Perhaps it is after a grub or other underground insect. It pushes its head and front feet into the excavation, then withdraws and scratches back the accumulated

At intervals the stream tumbles over waterworn boulders, falling into quiet pools where aquatic life abounds.

In contrast to the vibrant life in and near the waterfall, a snail creeps slowly up the stem of a plant.

earth. It does this several times. This goes on for ten minutes or more and I am increasingly mystified by its actions.

Skunks are fairly common in these mountains; they often amble into the light of our campfires on autumn nights. Once a large one came within two feet of my chair as I sat quietly enjoying the fire's warmth late at night. Closely related to the weasel and the mink, these large skunks are classified in the genus *Mephitis*, a Greek word meaning "poison gas," a most apt term, as can be verified by anyone who has ever had the temerity to molest one. As I watch the skunk digging quietly beside the stream, my mind drifts back down the years and I recall an experience of my boyhood friend, Charley. Charley and I attended a rural school in Montana and he was an enthusiastic trapper of muskrats and skunks and usually ran his trap lines each morning before reporting to school. On one such morning Charley found a large skunk in one of his traps but, unfortunately, the animal was caught by only one toe. As Charley approached, the skunk escaped and Charley, seeing his prize running away, made a flying tackle and grabbed the skunk. However, in the process the skunk discharged its stream of scent with great accuracy, hitting Charley squarely in the eyes. Undaunted, he dispatched the skunk and came on to school. As can easily be imagined, this brought on some discussion with the teacher, the outcome of which was that she forced him to sit out in the school-yard under a tree for the rest of the day. It was there that he did his lessons. In my mind's eye I can still see Charley seated under the spreading cottonwood tree, his feet propped up on its trunk and his book in his lap.

The skunk was no doubt the first creature to make use of chem-ical warfare; at least it is by far the best known. There is, however, a beetle—the bombardier beetle—that ejects a small cloud of irritat-ing gas from its anal extremity when alarmed. The odor of the skunk's musk is derived basically from chemical substances known as *mercaptans*. These are alcohol-like compounds containing sul-phur and other elements. Probably of all Nature's weapons of defense, the scent of the skunk is the most effective. As a result, a skunk has but little fear of man or any other creature; it goes about its business with an assurance and unconcern found in few other wild animals. Its scent is ejected under high pressure from twin glands beneath its tail and is golden-yellow in color. A num-ber of naturalists who have observed these jets of odoriferous musk at night state that they are phosphorescent or luminous. The scent also appear to be mildly toxic when coming in contact with the eyes or other tender membranes. My friend Charley, for instance, suffered considerable eye irritation from his encounter. Ernest Thompson Seton, the naturalist, records a case (1829) where sev-eral Indians lost their eyesight from encounters with skunks. If the scent enters the nose it causes choking and nausea, and some people have fainted as a result. On the skin it produces a burning sensation. The usual range of a skunk's "artillery" is not over five or six feet, but some observers have seen the musk ejected for ten or more feet.

Fortunately, the skunk is normally a docile creature when not molested. One night along Little River I followed behind a large skunk for several hundred feet, trying to get close enough for a photograph. Assisting me was my intrepid sister-in-law, Lucy Puckett, who held a flashlight. The skunk was uncooperative but not sufficiently alarmed to eject its scent, even though it raised its tail several times as if toying with the idea of doing so. At last we cornered the creature, foolishly perhaps, under an overhanging ledge and I crawled to within two feet of it in the cramped space to take a flash photo. It was not until later I realized how foolhardy I had been in my enthusiasm.

The American Indians were, of course, aware of the skunk and its unique means of defense. In each Indian nation there was a different name for it; the Hurons called it *Scangaresse*, and it is from this that the word "skunk" was evolved. The Crees and Ojibways called it *She-gawk*, a term that gave the city of Chicago its name. The Indian name means "Skunk-land."

I continued to watch the skunk busily digging in the earth beside the waters of Huskey Branch, still unable to discover its intentions. At this point a jay alights on a branch above the skunk and regards it with interest; no doubt it is the same bird that scolded from the rhododendrons during the sparrow-snake incident. The skunk, however, pays no attention to the bird and continues methodically digging, totally unaware that it has become the center of interest of at least two pairs of eyes.

Nothing happens for another minute or so, then the skunk retreats from its hole and appears to be slapping at something. For a moment I am unable to see the cause, then I notice a number of yellow jackets buzzing about, some of them attempting to sting the skunk. This solves the mystery; the animal has obviously been digging into the underground nest of a colony of yellow jackets in order to feed upon the plump grubs found in the paper cells. I can certainly sympathize with the skunk, since I only recently have had a painful experience with these same hot-tempered insects. While exploring a mountainside along Little River, I was pushing through the thick undergrowth when I felt hot, stabbing pains in my legs. Looking down, I found that I was standing almost on top of the entrance to a yellow jacket nest. Needless to say, I beat a hasty retreat, but not soon enough to have avoided a number of stings.

A skunk was discovered in the act of plundering the underground nest of yellow jackets to feed upon the larval insects.

In any case the skunk seems not to be bothered too much by the yellow jackets because it pushes back into the hole and begins pulling out fragments of the paper comb and devouring the larval insects contained therein.

Unnoticed during my preoccupation with the skunk, the afternoon sun has dropped down behind the mountain, leaving the stream and its environs in shadow. So, taking a last look at the skunk still feeding upon its ill-gotten booty, I turn about and retrace my steps along the stream and eventually down the precipitous path beside the waterfall.

Here along the river it is even darker than it had been above the waterfall; in fact, it is quite difficult to discern any detail in the deep forest along the river. Night is already approaching as I walk down the meandering road, yet when I look upward I can see the sun's last rays touching the top of the mountain beyond the river. This valley is, indeed, remote and secluded, and now, as if to emphasize this fact, the voice of a chuck-will's-widow drifts through the forest. On another day, perhaps I will return and penetrate even farther into the upper reaches of Huskey Branch, finding who knows what?

The trail through Cucumber Gap follows an old logging road. At one point it crosses Huskey Branch.

Chapter 6

CUCUMBER GAP

HALF A MILE OR so beyond the pool where the waters of Huskey Branch drop into the river there is a trail angling up the mountainside toward the right. It is an old logging road and so its slope is gentle and a walk along it not at all arduous. This, I must confess, is one of my reasons for often choosing it for biological explorations. It climbs up the side of the mountain, curving gently to the left, eventually crossing Huskey Branch, at that point a happy little brook bubbling along between mossy banks. Here and there the stream flows beneath fallen timber, disappearing, eventually, into a tangle of rhododendron and laurel. Farther down, its waters drop over precipitous ledges before joining the river, forming the attractive falls seen from the road.

I have explored most of this mountain stream but have never had the fortitude to follow its course through the dense jungle above the falls. Thus, there is perhaps a quarter of a mile of the stream that is unknown to me.

After crossing Huskey Branch, the trail continues on, still curving toward the left, and passing through deep woodland, eventually crossing Cucumber Gap near Burnt Mountain, then dropping down to Jakes Creek, a sizeable stream that joins the river at the lower end of Hidden Valley.

This pleasant pathway is shown on maps as the Cucumber Gap Trail, but I am somewhat confused as to why it was named thus.

On my first hike over it I saw numerous Fraser magnolias, known locally as "cucumber" trees, probably because of their large leaves. My assumption at that time was that the trail had been named for these trees. Later, when I again walked along the trail, I noticed large patches of Indian cucumber roots. I then decided that perhaps the trail had received its name from these plants. Now I am not certain as to the origin of the name, but after all, what's in a name? It is what one finds along the trail that is of interest.

Many times while traveling up and down the valley, I had glanced up the trail to Cucumber Gap, vowing that someday I would take the time to explore it properly; once or twice I had actually walked up it for a short distance, but that was all. Eventually, one spring day, I decided that the time had come to see what lay beyond view and so I started up the trail. It was with some trepidation that I walked away from my car and its safety, since there had been reports of a "bad" bear in the vicinity. However, I continued on, detouring around several marshy spots. At one point I left the trail to examine a flower, wading through tall growths of fern. All was quiet until suddenly near my feet there was a violent "explosion" as something burst out of the dense vegetation. Several seconds were required for me to regain my composure, after which it gradually dawned on me that I had flushed a ruffed grouse. This, of course, was not unusual nor the first time a similar thing had happened while I had been roaming the woods of Hidden Valley. Grouse have the habit of remaining quietly hidden in thick vegetation, taking flight only when almost stepped upon, always an upsetting experience when walking in the silent forest. In any case, the grouse flew away through the trees and disappeared.

Spring is the time to study ferns and there are tremendous numbers and kinds almost everywhere in the valley. This area, I think, should be called the "fern capital" of eastern United States; many kinds were no doubt evolved here and later spread westward and southward from this epicenter.

Earlier on this particular spring day I had examined and photo-

Walking ferns grow upon damp, moss-covered stones, their slender fronds stretching out and attaching themselves to new sites, where new fern plants are produced.

graphed walking ferns (*Camptosorus rhizophyllus*) growing on a damp outcropping of stone farther down the valley. These unusual ferns have lanceolate fronds, tapering to slender tips. Wherever one of the tips touches a damp surface a new plant is produced. Thus, in effect, they "walk" across the mossy stones upon which they grow, year after year spreading away from the point of origin. Some of the fronds, I noticed, had reached out beyond the stone, groping vainly for anchorage in the air.

Along the trail now there were several most attractive ferns. One of these was the maidenhair fern (*Adianatum pedatum*), unique in that its main stem divides, resulting in a distinctive form. Here they grew in abundance, their slender stems strung with tender leaflets, mostly arranged in one plane. Often these pretty ferns are found in the shade along the bases of the cliffs, always

The most attractive of all the ferns in the valley is the maidenhair, often found along the bases of rocky outcroppings.

where moisture is abundant. They occur in many places in the valley.

It was at this moment that I suddenly felt stinging sensations on my arms and the backs of my hands. Thinking that I had, perhaps, brushed against a stinging caterpillar, I looked among the ferns and other plants. I saw no caterpillars but I did find a nettle plant, its stems covered with needle-like spines. Knowing the cause of the itching, I was less concerned, although my hands and arms continued to itch for several hours. These troublesome plants are quite common in the valley, usually being found in damp places.

I came eventually to a place where the trail passed through a tangle of laurel, and paused to admire the surrounding flowers. Below me the mountainside was starred with thousands of white trilliums and white violets; it seemed almost as if they were lifting their blooms in supplication to the warm spring sun. Scattered here and there were bird's-foot violets, often called pansy violets, the largest and most attractive of all the local kinds. The blooms of one cluster, I was especially pleased to see, had upper petals of deep purple. These were "sports," or mutations, a relatively rare type. Almost everywhere I looked there were violets. A short

distance down the trail I noticed several with deep blue blooms
having varigated coloration. I was perplexed for a moment, then
recalled that this varigation is caused by a virus. I remembered, too,
that some of the most attractive colorations of camellias also result from virus infections.

Everywhere, for me, there were things of interest. In one place
there were several showy orchids and, in another, a puttyroot
orchid, its single, last-year's leaf dry and limp against the earth.

All was quiet in the forest except for the sound of the wind in
the trees and the call of an occasional bird. I walked slowly on,
reluctant to leave such breath-taking beauty. Scattered among the
living trees, some distance beyond the flowers, there were a number
of large stumps, all that remained of forest giants of long ago.
Beside one of these stumps I noticed a horse nettle plant with a
gayly striped beetle feeding on one of its leaves. When I captured

*Right: A puttyroot orchid grows beside a stump, its last-
year's leaf now dead. Mountain people once made a paste
from its bulbs for repairing broken pottery.*

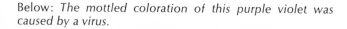

*Below: The mottled coloration of this purple violet was
caused by a virus.*

This potato beetle was feeding on a horse nettle. Upon its prothorax was an "H."

it, I was intrigued at noting that upon its prothorax was emblazoned an "H", my last initial. After studying the insect for a moment, however, I decided that it was merely a common Colorado potato beetle with an unusual marking. Evidently it had been feeding on the horse nettle, a member of the nightshade family and thus closely related to the beetle's normal food plant, the Irish potato. In any case, I hoped that this beetle was not an omen of evil portent. I knew that some people had seen significance in the W's on the backs of certain cicadas during World War II, thinking they stood for "War."

Finding of the brightly colored beetle took my thoughts off on another tangent. I recalled an entomological tale concerning one of the generals in the army of Napoleon I. It seems that this general, General Count Auguste DeJean, was an enthusiastic collector of beetles. He was, in fact, a well-known authority on these insects. During the battle of Alcanizas, General DeJean saw a rare, brilliantly colored beetle perched on a flower beside a small stream he was about to cross. He quickly jumped from his horse and captured the beetle, pinning it to the inside of his helmet. During the ensuing battle his helmet was pierced by a ball but the general was unharmed and later found his precious beetle still intact. It turned out to be *Cebrio ustulata*, a wasplike member of the family Cebrionidae. Incidentally, General DeJean won the battle, taking many prisoners. In pondering over this incident, I wondered if I would have been as concerned, as was the general, about a small beetle during the heat of an important engagement. All I had to worry about was the possibility of meeting a mean bear!

There were, however, other dangers, as I was shortly to discover. I had pushed through a mass of dense vegetation and was in the act of stepping over a rotten log when some sixth sense warned me to look at the ground on the opposite side. There, resting quietly upon the leafy carpet, was a large copperhead. As I stood, uncertain what to do about it, the snake crawled leisurely away through the ferns and disappeared from view, while I retreated to the safety of the trail. Needless to say, I was careful to look before I stepped during the rest of the day.

There are, of course, many dangers in these mountains. In examining leaves for insects, my hand inadvertently brushed against a spiny slug caterpillar (*Euclea*), causing an intense burning sensation that lasted for some time. The caterpillar had been hiding on the underside of an oak leaf and was about an inch in length, varie-

Copperhead moccasins are fairly common along the trail. Sometimes they rest quietly upon the leafy ground or in the dense vegetation.

The waters of Huskey Branch drop down through a mossy glen from the higher mountainside.

gated in color and rather attractive, I had to admit. On previous jaunts I had frequently seen these poisonous little caterpillars, almost always on the under surfaces of leaves.

Some distance beyond the stump where I had seen the potato beetle I came to Huskey Branch and the point where the trail crossed it. By stepping on slippery stones I was able to get across, and I then sat down on a boulder to contemplate the peaceful scene. Above me the water tumbled down over mossy boulders, falling into clear pools with pleasant, rippling sounds. The stream continued on down the mountainside, disappearing eventually into a dense growth of laurel known locally as a "hell." It was near that point that I noticed a place where the earth had been freshly disturbed. This seemed a bit unusual, so I walked over to investigate. Apparently some animal or animals had been digging or rooting in the soil to depths of nearly a foot and covering a considerable area. This, at the moment, had me puzzled, then I remembered that there had been reports of wild Russian boars in the vicinity, and the signs I found reminded me of the rooting activities of the hogs I had previously seen in the Cherokee National Forest along the Tellico River. I knew that authorities here in the Park were much concerned about the imminent invasion of the area by these animals which they had been attempting to prevent by means of trapping operations down in Cades Cove.

These exotic swine were introduced into the mountains of North Carolina in 1912 as a game animal on a large shooting preserve in the Snowbird Range, a remote area about twenty-five miles north of Robbinsville, North Carolina. This is only a short distance from a high bare area or "bald" known as Stratton Meadow which I once visited.

In the process of preparing this hunting preserve, twenty-five tons of barbed wire were laboriously hauled into the mountains by wagons, the boar enclosure consisting of 6,000 acres. A club-house and a caretaker's cottage were also built. In addition to European boars, the animals introduced included fourteen elk, six Colorado mule deer, several buffalo, thirty-five bears (including nine large Russian bears), two hundred wild turkeys, and about fifteen wild European boars. To complete the stocking of this Noah's Ark, 150 sheep, and several hundred tame turkeys were added as food for the predators.

The remarkable thing was that only the wild boars found the

Stinging nettles (Laportea canadensis) are quite common along the trail through Cucumber Gap. Contact with the skin causes a severe burning sensation.

mountains to their liking and thrived in spite of hunting by dogs and men. The original shipment had consisted of ten or eleven sows and three male boars, evidence indicating that they had been trapped in the Ural Mountains of Russia. It is for that reason that they are known locally as "Russian" boars.

Eventually the promoter of the hunting preserve lost interest in the project and a wild boar hunt was held with the alleged purpose of destroying the animals. Men and dogs converged on the fenced enclosure and the hunt began. However, it did not go as planned; when it was over, only two wild boars had been killed, while half a dozen dogs lay dead. The intrepid hunters had spent most of their time in the safety of trees. Meanwhile, the boars escaped from the fenced area and extended their domain away through the mountains, adapting themselves remarkably well to the remote parts of the wilderness area, everywhere clothed with dense tangles of rhododendron and laurel. In the process of extending their range, the Russian boars interbred with domestic hogs, resulting in a strain retaining the tougher and meaner characteristics of each.

An adult boar—both males and females are called "boars"—may weigh up to five hundred pounds, though the average is perhaps half that. As is frequent in wild pigs in many parts of the world, their young have longitudinal stripes that disappear after about six months. Both males and females have greatly elongated canine teeth called "tusks," located in both upper and lower jaws. These tusks rub against each other and are thus "self-sharpening." The boars have few enemies, since an adult, especially when cornered, may be extremely vicious.

These wild hogs range widely, destroying vegetation by their deep-rooting habits. They also devour both eggs and young of ground-nesting birds. Altogether, they are an undesirable addition to the native fauna of the mountains.

Here, beside the Cucumber Gap Trail, I studied the disturbed earth, attempting to decide whether or not it indicated the presence of wild boars. There were no tracks on the leaf-littered floor of

At one place along the trail were many ground pines or Lycopodiums. The ancient ancestor of these club mosses grew to tree size.

the forest and, at last, I gave up, uncertain as to what had dug up the ground. It may or may not have been wild boars.

Just beyond the Huskey Branch crossing, the trail rounded a high bank and angled off toward the left, continuing straight through the forest in a gentle ascent. Here grew a mixed stand of trees, one of the more attractive and conspicuous being a Fraser magnolia, the sun shining down through its great leaves, bright green and translucent. It reminded me of the big-leaf magnolia so common in the Deep South. Unknown to many, there are at least nine different kinds of magnolias in southeastern United States. The one here—*Magnolia fraseri*—was named for John Fraser, a Scottish plant collector who visited the Carolinas nearly two hundred years ago.

Beneath one of the larger magnolias the ground was covered

with ground pines, looking like miniature pine trees, and averaging about five inches high. As a matter of fact, many people think that these are actually young pines, but the truth is that the attractive little "trees" are a type of club moss, *Lycopodium*, arising from rhizomes creeping along beneath the leaf-litter. These strange little plants have a long story to tell; their remote ancestors grew to tree size. Some of them, such as *Lepidodendron* or "scale trees," resembled large palms with bushy crowns. However, the scale trees are gone, relegated by evolutionary change to the scrap heap of discontinued models; only their fossils remain. Yet, here along the trail their descendants grow as small mosses, now dominated by great trees more recently evolved.

Many of the ground pines beside the trail bore brownish spore-cones or *strobili*, the sites where their reproductive spores are produced. In the same damp area I found specimens of still another kind of club moss, this one with unbranched stems, looking more like ordinary moss plants but of larger size. As I looked at these odd little plants, I was reminded of the long history of the plant life of these mountains; the club mosses were evolved more than 300 million years ago, long before the first flowering plants appeared.

Always, it seems, I am intrigued by the origins of plant and animal names. The botanical name of the club mosses is *Lycopodium*, derived from two Greek words, *lykos*, meaning "wolf" and *pous* or *podos*, meaning "foot." When combined, the names mean "wolf's foot," which seemed quite farfetched, since I could see no resemblance.

Unnoticed while I had been studying and photographing the ground pines, a heavy cloud bank had built up in the west and I was surprised when drops of rain began falling. Hastily, I retreated to the scant shelter offered by a nearby hemlock and quickly covered my bag of photo equipment as best I could with my jacket. The rain trickled down the hemlock twigs and dripped from the tips of its needles, splashing upon the fallen leaves and collecting like gems on the mosses. To me, the falling rain created

The spurred violet is easily identified by its pale violet coloration and the long spur. It grows along the trail.

discomfort but the surrounding plants, if I may be permitted to use the term, seemed "happy." I could almost see them absorbing the moisture falling upon their leaves. The ferns held their fronds a little more erect. Some kinds of animal life, too, reveled in the falling moisture. Snails emerged from their coiled shells and moved slowly over the decaying leaves, while, on a fallen log, a large slug was aroused to activity, creeping phlegmatically across the mossy carpet growing there. It left a shining slime trail behind as it moved along.

But not all the forest inhabitants found pleasure in the shower. Chipmunks, usually active everywhere while the sun had been shining, disappeared, retreating, I suppose, to underground dens. Beneath a leaf a swallowtail butterfly hung suspended by its legs, its wings tightly folded. A creature of the sun, it would remain

Spleenwort ferns are in abundance, often growing out of crevices between stones.

quiet until the shower had passed. Meanwhile, its scaly wings were shedding the water like the feathers of a bird.

In time the rain cloud drifted away toward the east and the sun reappeared. The swallowtail flexed its great wings and flew off down the trail and, from high in the trees, I heard the call of a bird. Although water was still dripping from the hemlock, it was no longer raining, so I left my shelter and walked on.

The trail, at this point, was bordered by ferns of many kinds and upon the boulders were rock-cap ferns, sometimes known as polypody ferns. Along the bases of the same boulders I noticed delicate spleenworts (*Asplenium*). Here and there grew a number of rattlesnake ferns (*Botrychium*), each with its fertile frond thrust straight upward above its feathery leaf. Why, I wondered, does this attractive fern have a name that has an unpleasant connotation remindful of danger?

Some of the ferns were just uncoiling their fiddleheads, unrolling their filmy leaves in the manner so characteristic of these plants. I examined some of them closely, noticing how the fiddleheads expanded in geometric spirals like the shells of snails, but always with their stems on the outside of the symmetrical coils.

The time was passing so I hurried on, coming eventually to a large patch of Indian cucumber roots (*Medeola*) with their leaves attached around their stems like the spokes of a wheel, and with small, greenish flowers. I had never seen so many in one place and so was especially interested. The name cucumber root, as might be supposed, was given the plant because its rootstock tastes something like a cucumber and so was no doubt eaten by the Indians. It belongs to the lily family and its generic name, Medeola, is derived from Medea, the evil sorceress of ancient Greek mythology who sent her rival a wedding gown anointed with deadly drugs in which she was burned to death. However, in spite of this rather gruesome recollection of the origin of their botanical name, I found the plants most attractive and of characteristic form. When,

Rattlesnake ferns are easily identified. The spore-bearing frond grows upward from the fern's center.

Fern fiddleheads rise beside a stump like large green question marks.

Opposite top: Indian cucumber roots have small, greenish flowers. In autumn, they bear clusters of purple berries.

Bottom: Oil nut trees, with their greenish flower clusters, are often seen in spring. The tree lives as a parasite on the roots of other trees.

in autumn, I again passed along the trail, their leaves had turned brilliant red and their blooms had been replaced by purple berries.

With the abundant plant life as closely associated and competitive as it is in these mountains, it is not at all remarkable that

Opposite: *Lacking green chlorophyll and pale in color, cancerroots live as parasites on the roots of other plants, including those of violets.*

The nuts of the oil nut tree are rich in poisonous oil.

a number of plants live as parasites upon others. In more arid regions having sparse vegetation there are few parasitic plants, but as one travels southward into warmer, more humid, climates one finds many plants that live by drawing their nourishment from others. In the tropics I have seen large sandalwood trees that lived by joining their roots to those of other trees in order to steal their nourishment. Along the Cucumber Gap Trail I found a local member of the sandalwood family (Santalaceae). This was the oil nut (*Pyrularia pubera*), not an uncommon tree in these mountains. The one beside the trail was a little over six feet tall, of spreading form, and bore many spikes of small, greenish flowers. Under casual observation there seemed nothing remarkable about this shrubby little tree but, knowing that it belonged to the sandalwood family, I was aware that it lived as a parasite upon the roots of rhododendrons and, perhaps, other plants or trees. In autumn it bears nuts about the size of a hickory nut, of ovoid form and rich in oil.

On a high bank, a little farther down the trail, I found two other parasitic plants, both members of the broomrape family (Orobanchaceae). To give due credit to the oil nut tree, it does

contain green chlorophyll and so can maufacture its own food, absorbing chiefly minerals from the roots of its hosts. The broom-rapes, by contrast, are devoid of chlorophyll, depending entirely upon what they can steal from the roots of other plants. One of the parasitic plants I found was cancerroot. There was a large group of their attractive little blooms, rising perhaps four inches above the earth, all facing in the same direction. The other broom-rape was squawroot, looking like fleshy, white stalks of asparagus or, perhaps, yellowish spruce cones, set in the ground. These plants, I knew, lived as parasites on the roots of the nearby oaks and, perhaps, other trees.

Along the trail I noticed ants of several kinds. In these mountains more than sixty different species have been recorded. However, the higher one climbs the fewer ants are found, until, at 6,000 feet, there are only two kinds. It is obvious that climate has a profound influence on their distribution. In Hidden Valley, I would guess that there are about thirty species, many of which I have observed during my rambles.

The squawroot, yellowish in color, derives its nourishment from the roots of oaks and other trees.

Stinging caterpillars cling to the undersides of leaves. Mottled green in color, their spines contain poison.

A common ant along the trail is the large, black carpenter ant (*Camponotus*) and on this first hike I noticed a number of them running across the path. These ants always seem to be in a hurry. Being predatory by habit, they are constantly in search of small insects or other game which they capture and carry back to their nests in stumps or rotten logs. I seated myself upon a boulder and watched some of them carrying captured insects toward a log where no doubt they had a nest. One worker ant had captured a small moth and I was surprised at the amount of physical exertion required to carry it. The ant tugged the moth, which was considerably larger than itself, over and under dead leaves, sometimes dropping it, then retrieving it, always making gradual progress toward the rotten log. At one point a stone, perhaps six inches in diameter, was encountered. Foolishly, it seemed to me, the ant hauled the inactive moth up the side of the stone. In the process it dropped the moth several times, but it was persistent and at last succeeded in dragging its prize across the top of the stone and down its opposite side. While watching the toiling ant, I was tempted several times to give it a helping hand but decided that my aid would not be appreciated. To the ant there was apparently but one direction to carry the moth—directly toward its nest. From my vantage point, however, it seemed obvious that it would have been easier for it to have detoured around the stone instead of going over it. In any case, I decided that this was the ant's problem, not mine, and so I was able to suppress my intentions as a good Samaritan. After a few minutes the ant was successful in dragging the moth into one of the nest entrances and I looked about to see if any other ants needed my "help." Some distance away another ant apparently could have used some sort of aid. It had found or captured a butterfly and was attempting to carry one of the large wings back to the nest. This unwieldy object was difficult for the ant to carry; it was continually becoming lodged among leaves and the stems of grasses, and the ant's progress seemed so hopeless that soon I lost interest in its struggles.

At this point I noticed some sawdust falling from the end of a dead twig beside the trail, and when I moved closer to investigate, I found that one of the black carpenter ants was busily excavating a tunnel in the twig. Every few seconds the ant would appear in the end of the twig, discharge a load of fine sawdust, then disappear. It was working diligently and a small pile of dust had accumulated on the ground about a foot below. I was at a loss to know why the ant, obviously a worker, was tunneling into the twig. If it had been a queen I could have understood the reason for her work; she would have been in the process of founding a new colony. Without doubt, this ant was a member of the nearby colony and so I could think of no good reason for its activity. But then, the habits of ants are not always logical, at least to the human mind.

With my machete I cut into the carpenter ants' nest in the rotting log, ruthlessly exposing their living quarters, a desecration done in the interest of science. The disturbed workers ran about, some of them vainly attempting to defend their grublike larvae which some of the workers were carrying deeper into the honeycombed wood. I had about decided that nothing more of interest

Living in the nests of black carpenter ants are tiny crickets. Tolerated by the ants, treated almost like pets, these myrmecophilous crickets are found nowhere else.

was to be seen in the nest when I detected a tiny insect hopping about among the milling throng of ants. Looking more closely, I saw that it was a cricket, and I was fortunate in capturing it in a vial. This little cricket turned out to be an ant-loving cricket or *myrmecophile*. These little insects—they are smaller than the ants—dwell in apparent safety with the pugnacious ants, tolerated by them. Living within the ants' nest, they feed upon secretions from the ants' bodies, always avoiding their sharp jaws. These crickets are wingless but have large hind legs, well fitted for jumping, and I suspect that their agility enables them to survive in their dangerous habitat. They are found nowhere except in the nests of ants, a life to which they have become adapted over millions of years. Myremecophilous crickets of other kinds dwell in the nests of other species of ants, but the same kind of cricket is always found associated with the same kind of ant. Those living with smaller ants are also small, matching, in a comparative way, the size of their hosts.

The place where I had found the carpenter ant nest was located in a glade, somewhat dryer than most of the surrounding forest, and so I was not surprised to see other ants crawling about. As I walked through the sparse vegetation I noted a number of ants fighting. This battle was taking place near a large, flat stone and, looking closely, I saw that there was an ant nest beneath it, as evidenced by the quantities of plant debris piled about its edges. Everywhere on the surrounding ground ants were in combat, their jaws and legs interlocked. It gradually dawned on me that I was witnessing a raid by slave-making ants.

The nest beneath the stone, I discovered, was that of black formica ants (*Formica fusca*) and a large number of red slave-hunters (*Formica sanguinea*) had attacked it and were carrying off pupal ants enclosed in cocoons. The black ants were attempting to repel them; however, many of the slave-hunters had entered the nest and were successfully carrying off their booty, while others were battling the owners.

For perhaps half an hour I squatted down by the stone, watching the little drama, fascinated by what I was seeing. It was only the second time in my career as an entomologist that I had been privileged to see one of these slave raids, and so I followed some of the red hunters with interest as they carried their stolen pupae across the leaf-littered ground to their nest under another stone, perhaps twenty feet distant. Hurrying along with their burdens, they struggled over the uneven ground, sometimes dropping the cocoons, then picking them up and continuing on, always in a great hurry.

The slave-master relationship between these closely related kinds of ants has been studied in detail by many formicologists, including William Morton Wheeler and A. C. Cole, the latter having studied the ants of the Great Smoky Mountains for many years. Once the captured pupae in their cocoons are safely in the red formicas' nest, they are cared for as if they were their own young. When they emerge as adults, they serve their masters by doing much of the work in the nest. The red formicas are quite able to get along without their slaves, but the presence of the additional labor force is undoubtedly of great aid to them.

Colobopsis *ants nest in tunnels in white ash twigs. The soldiers have plug-shaped heads which they use as stoppers to close the entrances to the nests.*

With the subject of ants still on my mind, I examined the twigs of a white ash to determine, if possible, whether or not they contained the nests of trap-door ants (*Colobopsis*). At another place in the valley I had previously found these remarkable ants nesting in white ash, so I suspected that they might occur here along the trail. After examining a few twigs, I was pleased at finding some with typical entrance holes that I was certain led into *Colobopsis* nests. I split one of the twigs open and was rewarded by the sight of several workers, some white larvae and, most interesting of all, three soldiers. It is these latter ants that are most remarkable both in form and habits. I captured one in my fingers and looked at it with my hand lens, especially its peculiar, plug-shaped head. Nowhere in nature, as far as I know, is any similar adaptation found. It is the function of these soldiers to close the round entrances to the nest tunnels in twigs, using their plug-shaped heads as stoppers. Their cylindrical heads fit snugly in the holes, with their flattened fronts toward the outside, thus forming living doors to the nests and preventing enemies from entering.

I examined a number of other nest entrances in the ash and, in each case, saw a soldier's head plugging it. I also noticed workers returning from forays out on the tree. When one of these approached the entrance, the soldier "on duty" obligingly backed away, allowing it to enter, then "shut the door" again. How, I wondered, did the soldier recognize the worker as a member of its own colony? The soldier's eyes are so placed that it cannot see outside; the "password" is evidently the worker's odor. I found another *Colobopsis* nest in a nearby ash and placed an ant from it on the twig near the entrance to the colony I had been observing. It crawled about for a few minutes, then attempted to enter the nest. The soldier on duty refused to open the gate and the "foreign" ant eventually lost interest and crawled away.

So engrossed had I been in my observations of ants that I had forgotten the time. I now found that dusk was approaching, so I hurried back down the trail. I certainly had no desire to let darkness overtake me in this remote area. By the time I had reached my car down along the river, it was completely dark and the valley was silent except for the persistent voices of frogs and of an owl calling from far up the river. Hidden Valley was, indeed, a lonely place at night. I drove down the deserted road, passing the falls and then continuing across a level, forested area. Suddenly, far ahead, I saw some animal, barely visible in the headlights. Driving on, I saw it to be a large skunk, which, as are the habits of these animals, was unhurried by the sound of the motor. As if daring me to hit it, it finally trotted up the embankment and disappeared in a dense growth of ferns.

At this point I remembered that I had with me a "varmint" call, a small whistle-like call similar to those used to call crows and ducks. This little device, it was alleged by the manufacturer, would call up all sorts of varmints. It was supposed to operate on the theory that since it sounded like the cry of a wounded rabbit, it would attract predators. I rolled down the car window and turned off the lights, finding myself suddenly enveloped in silence so profound as to be almost frightening. I felt almost as alone as if

In a beam of light, the white flower clusters of dog-hobble seem to glow with fluorescence.

I had been on the moon. I blew a tentative squeak on the varmint call, then waited. I heard nothing except that of the wind in the trees. I gave another call and again waited. Nothing. By this time I was feeling very "creepy," so turned on the headlights. I had, indeed, attracted an animal! In the beams of the headlights I saw a white-footed mouse running across the road.

I am sure that these varmint calls, more properly known as "predator" calls, are effective if one has the patience and is in an area where such animals as foxes and bobcats are abundant. In any case, I drove on, seeing no other wild creatures. The pendant clusters of dog-hobble blooms along the way seemed to glow in the car's headlights as if they contained tiny electric bulbs. There were other white flowers along the way, but none seemed as fluorescent as did those of the dog-hobble. These forests, interesting and beautiful by day, are, indeed, mysterious and strange by night. I definitely prefer the day.

Chapter 7

AUTUMN ON THE TRAIL

A NATURALIST—I have forgotten who—was once asked by a friend how he had spent the summer. "Well," the naturalist replied, "I started exploring my back yard. However, I got only about halfway across it."

The gentleman was probably being facetious, yet it is true that a biologist revels in minutiae, in small details of the world around him. Mine is, perhaps, a selfish interest; I like to wander along a trail or explore a forest, seeing few other humans and certainly no discarded beer cans. In short, I prefer that wild places remain wild. Yet, I know that for such places to be preserved in their primitive beauty there must be public interest and support in a world where population pressures are straining at their bonds. Hopefully, Hidden Valley and its trails will remain a wilderness, with no paved highway passing through it, carrying high-speed traffic, drowning out the pleasant sounds of the river and the songs of the woodland birds.

The valley is a microcosm, a relatively small, isolated area where a biologist could easily spend a lifetime, with each day, each hike, opening up new vistas, new wonders to be investigated. On visits to the same spot, even on successive days, I always find new things. Fungi grow overnight upon rotten logs; flowers appear where

The greenish flowers of Indian cucumber roots are replaced in fall by purple and white berries.

131

yesterday there were none. Even more diverse are the changes brought about by the seasons.

The trail leading out of the valley through Cucumber Gap had been fascinating in spring when the ground was starred with myriads of white trilliums and violets, the leaves of the trees freshly opened, and the dogwoods arrayed in snowy blooms. I hoped that it would be equally interesting in autumn.

It was a bright October day when I again passed along the trail, unconscious, as usual, of any real objective. Many of the trees had already lost their leaves and they now lay upon the ground like a multihued carpet. In spring the massed wall of foliage had obscured the more distant views, but now the mountainside stretched away, flowing upward and over the crest, the stark forms of the distant trees outlined sharply against the skyline. Great stony out-croppings and fallen logs were revealed, unseen during my previous hike. A new facet of the forest's personality was in view. As I walked along, the fern-bordered trail looked familiar, yet changes were everywhere evident and I experienced the pleasant illusion of exploring a new and unknown place.

The flowers of spring had disappeared. The leaves of the trilliums hung bedraggled and sad from limp stalks; on only a few were there ovoid berries. The Indian cucumber roots were recognizable by their deep purple fruit. In one place, beside a log, was a stalk topped by a cluster of scarlet berries, the fruit of the jack-in-the-pulpit. The plump berries added a bright spot of color to the autumn landscape. Scattered through the forest were dogwood trees, their leaves now tinted scarlet, each twig tipped with bunches of red berries.

Seasonal changes had transformed the mountainside; it seemed as if a curtain had been swept aside and one vast stage setting replaced by another of equal beauty. Some of the characters were the same, others were different, but each was playing its role in the pageantry of the seasons. Where, in spring, there had been violets, orchids, and lilies, there were now mostly goldenrods and asters. In general, members of the composite (daisy) family were

In autumn, clusters of glossy-red berries appear on jack-in-the-pulpits.

Below: Scarlet berries appear on the twigs of the dogwood trees.

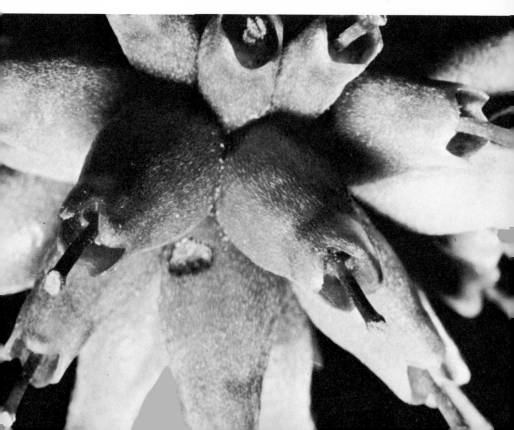

Beside the trail a swallowtail butterfly sips nectar from the bloom of a trumpet flower.

now in ascendance, dominating the scene. Each flower, I reflected, has its day in the sun, its cue to blooming triggered by the shortening of the days in a cycle almost as old as these ancient mountains.

Many of the plants and trees along the trail now bore seed pods or fruit instead of blooms, the end results of all the complex physiological process of spring and summer. Winter would soon arrive, but each plant, in its own way, had been preparing for it. Some had stored up food in their roots or stems, others had produced, or were producing, seeds as a means of bridging the seasons.

How nicely Nature balances her books, keeping those plants—and animals—that are fitted to survive, dicsarding those that cannot adapt themselves. Seeds are packages of life, enclosed in hard coats, capable of living through adverse times, through periods of drought and cold. We admire a flower for its beauty but, in Nature's scheme, the flower is merely one small step leading ultimately to the production of seeds, the plant's complicated method of assuring the continuation of its kind.

In spring I had seen a number of oil nut trees along the trail, all bearing insignificant flowers. On some of these same trees I now saw large, nutlike fruit. When cut open, their contents were quite oily, the reason, of course, for the tree's name. Sometimes these trees are also called mountain coconuts or buffalo nuts. In any case, the oil is ill-scented and poisonous.

Near the oil nut grew a sweet shrub (*Calycanthus floridus*). In spring it had been adorned with maroon flowers, sweet-scented, with petals and sepals of the same color and straplike in form. Now from its twigs were suspended seeds enclosed in attractive, urn-

In spring, the twigs of mountain silverbells are festooned with white bell-like flowers. In autumn, these are replaced by winged seeds.

The characteristic seed pods of dog-hobble hang in clusters beneath the twigs.

shaped receptacles. There were sprigs fallen from nearby sourwood trees, each one strung with rows of pendant seed pods. In one place there were several winged seeds from a silverbell towering high above the trail. In spring I had gazed in admiration at the tree's rows of silvery, bell-like flowers. The dog-hobble beside the trail had replaced its creamy-white blooms with rows of brown seed pods of characteristic form. Everywhere, the trees of the forest had produced their nuts and seeds, dropping them indiscriminately upon the ground. They lay scattered about, some showing the tooth marks of squirrels and chipmunks. Beneath a large tulip poplar were many of its burlike cones with linear scales. Most attractive were the small hemlock cones seen among the fallen leaves, while beneath a large white pine were a number of its elongate cones, their scales now widely spread and the seeds gone.

One of autumn's most elegant seed-bearing shrubs was the strawberry bush (*Euonymus*), so-called because of its crimson, three- to five-lobed capsules. Now opened, each capsule had glossy-red seeds suspended from it. These seed capsules were covered with short, wartlike spines, a characteristic which gave the shrub its name. Sometimes they are also called "hearts-a-burstin'" because of their fanciful resemblance to broken hearts.

Along Huskey Branch, near the point where the trail crosses it, I had, in spring, noticed a growth of jewelweeds or touch-me-nots, a common herb found almost everywhere in the valley. They were now adorned with elongate seed pods and, since I have never been able to resist the temptation, I spent several minutes touching them, intrigued by the little explosives pops they made as they split open, tossing their seeds away through the air. What a marvelous mechanism this is for seed dispersal! Apparently, tension builds up as the pods mature, causing them to explode at the slightest touch. Yet the witch hazel has an even better mechanism. Its hard, streamlined seeds are located between tough, liplike sections of their pods.

Left: The scarlet seeds of the strawberry bush, or wahoo, hang suspended beneath an open bur which is bright red and covered with short spines. Right: The hard seeds of witch hazel are compressed between fiberous lips. They are snapped away with great force.

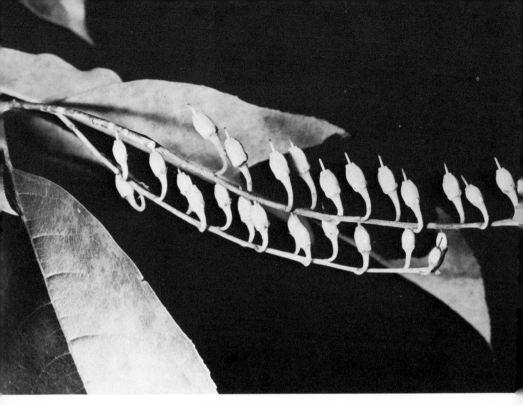

Sourwood seed pods stand upright along the twigs like rows of small candles.

As in the case of jewelweed, tension builds up, but it is much greater, eventually shooting the seeds away for a considerable distance.

The afternoon sun was now slanting down through the forest, and against the dark tree trunks floated lacy filaments of silvery spider silk. They undulated gently in the slight breeze, some drifting above the stream. Attached to each filmy mass of silk there was a tiny spiderling, obtaining a free lift, buoyed up by its silken parachute. Also floating through the air were the parachutes of dandelions and other autumn composites. To each delicately formed parachute was attached a seed being transported to a new and perhaps favorable place for growth. The forest was silent, yet many things were taking place around me in the orderly preparation for another year of growth and fruition.

Beside the stream was a large, flat boulder, the same one upon which I had rested in spring. I sat down upon it and surveyed my surroundings. The stream still rippled down from pool to pool, the moss along its margins was still green; only the ferns looked weatherworn. Some distance away a groundhog lay half-asleep upon a high boulder, its fat body pressed upon the sun-warmed surface. He, too, had made preparation for the coming winter without knowing that he was doing so. He had a snug den beneath his boulder and, during summer, had grown fat upon the lush vegetation. In the not too distant future he would stop eating but continue to doze each day in the autumn sun. About two weeks later, when frost lay white upon the ground, he would fail to appear for his usual siesta. Curled up in his den, he would be asleep, his temperature lowered, and heartbeat and breathing slowed down to the bare minimum needed to sustain life. Hibernation, from the Latin *hibernus*, means "winter," but the groundhog would know nothing of winter or its cold; in spring he would have no memory of that season. In spring the groundhog would revive and crawl from his den. He would nonchalantly begin nibbling upon the succulent spring vegetation, feeling, perhaps, deep within his consciousness, the reproductive urges of his kind. At the present moment, however, he lay quietly upon his boulder, unconcerned that autumn was speeding toward its inevitable end.

While I had been ruminating upon the ways of groundhogs, a pert chipmunk had run down the length of a log and now paused, regarding me with bright eyes. Unlike the groundhog, it was alert and vivacious, ready at any instant to dart away should I show the slightest aggressive tendency. I rummaged in my camera case for remains of my lunch and tossed a crust of bread in the chipmunk's direction. It quickly disappeared over the opposite side of the log with a flick of its tail, then reappeared, its inquisitive nature dominating its alarm. I now saw that its cheek pouches were already stuffed with food, probably seeds or nuts, but it at once crammed the bread in, puffing them out even farther. Here, I realized, was yet another approach to winter survival. The groundhog had prepared himself by storing fat in his body; the chip-

Chipmunk poses.

munk, by contrast, was laying away a store of food in its den. Who can say which is the better method? During the winter months the chipmunk would revive now and then to nibble at its store of seeds and nuts, then lapse into sleep again. A few minutes after it had hurried away to its den I noticed another one on the opposite side of the stream. This one was carrying a bundle of dry leaves, which seemed a strange behavior until I realized that this was lining for its winter nest.

While watching the chipmunks, it occurred to me that it would be interesting to trap a local chipmunk and take it home with me when I left the mountains. Unfortunately, there are no chipmunks near my home. There I would keep it in a cage, observing it through the winter. Accordingly, I later live-trapped a chipmunk and, upon arrival home, placed it in a sizeable cage in an unheated workshop adjacent to my studio. I fed it nuts and other appropriate foods, and it eventually built itself a nest. To all appearances, it settled down to a happy life in spite of the fact that, when transferring it from trap to cage, the tip of its tail had been accidentally severed. As a result, it was known from that time on as Bobby.

When the weather turned cold, Bobby disappeared into his nest, but I noticed that food placed in the cage usually disappeared, proof that he occasionally emerged to eat. At last, in mid-January, my curiosity got the better of me and so I removed the pieces of bark beneath which he had made his nest. Bobby was curled up tightly and I could see only occasional breathing movements. Seemingly, his metabolic processes had slowed down considerably.

A chipmunk lies curled up in the deep sleep of hibernation. Sometimes during winter it revives, feeds, then goes to sleep again.

I prodded him with my finger, but he did not stir. However, I could see that he did begin breathing very rapidly and, after a few minutes, crawled out of his nest and ran about the cage. Apparently his increased rate of breathing, upon being disturbed, had supplied abundant oxygen to his body, fanning his life-fires into normal function and, as a result, he had revived. Several times during the remainder of the winter I interrupted Bobby's winter sleep, always with the same results.

When the time arrived to return to the mountains, my wife and I placed Bobby in a small wire cage and took him along, liberating him at the same spot where he had been captured the previous fall. I then realized that, from our standpoint at least, the severing of his tail had been most fortunate; it furnished us with an easy means of identification.

Upon his release, Bobby disappeared into a mass of vines and was not seen again until the following day. I had about despaired of ever seeing him again when out of the vines darted a chipmunk. A second glance, however, showed that it had a long tail and so it was not Bobby. However, the next second another chipmunk appeared, this one with a short tail and we knew that it was indeed Bobby! The two chipmunks chased each other about for several minutes while we watched, elated that Bobby had apparently found a friend. During the subsequent weeks we placed food upon the ground near the tangle of vine and he and his friend obligingly came out and fed upon it. Bobby was much tamer than the "wild" chipmunk, having become accustomed to me, I suppose, during the winter. Actually, while in his cage, he had never shown any indication of being any less afraid of me. As far as I know, Bobby still lives in his native mountains, and I often wonder if he has dim memories of his sojourn in the Deep South. Probably not.

And so each plant and animal in the valley, in its own fashion, was making ready for winter. Earlier in the day I had noticed several woolly bear caterpillars humping across the trail as if in a great hurry. I thought of the old superstition regarding these insects: if the rust-red band around a woolly bear's middle is wide, a long winter may be expected; a narrow band indicates a short winter. The old mountaineers swore by this method of weather forecasting, but I have examined woolly bears for years and never noticed any difference in the widths of their bands; perhaps all winters in the mountains are long and severe!

In any case, the woolly bears, too, were getting ready for winter. Each one must find an insulated place beneath a rotten log or perhaps inside a stump. There, fairly well protected, it curls up into a ball and remains inactive until spring when it will revive, feed for awhile, then spin its cocoon.

An Aesop's fable tells the story of the industrious ant and the improvident grasshopper, but the fact is that the grasshopper is not so improvident after all. Along the trail, in an open area, I noticed a female laying her eggs. Her abdomen was almost completely buried and I could see the circular marks made by her feet as she had "screwed" her abdomen into the sandy soil for the laying of eggs. Thus, even the grasshopper was making provision for the future.

I continued on up the trail, stopping now and then to examine flowers and mushrooms along the way. The autumn sun was warm, and an occasional breeze stirred the foliage of the trees causing showers of dead leaves to fall. Beside the trail, in one place, grew a clump of goldenrods, their spikes glowing golden in a beam of sunlight. As I glanced at the mass of flowers, my eyes were attracted by a monarch butterfly sipping nectar from them. Momentarily the butterfly probed for nectar, then fluttered away through the woods. However, as I looked upward, I was surprised at seeing a large flock of monarchs flying through the forest. There were several dozen of them, all flying in a southerly direction as if following a compass course. The group fluttered across a forest glade, their rust-red forms sharply outlined in the sun, and disappeared among the trees. Soon, another group appeared, going in the same direction as the first. It now dawned on me that I was seeing one of the autumn migrations of these well-known butterflies that migrate northward in spring, often traveling all the way to Canada

Even the grasshopper prepares for winter. Here, a female bores her abdomen into a sandy spot to lay her eggs. She will die in winter's cold, but her young will hatch in spring.

and engaging in return flights in autumn. I recalled, too, that one monarch that had been tagged in Ontario, Canada, on September 18, 1957, was captured by a small Mexican boy on January 25, 1958, at Estacion Catorce, Mexico. This was a airline distance of approximately 1,870 miles, certainly a remarkable feat for a fragile insect.

During subsequent days additional flights of monarchs were observed in the general area. Down in Hidden Valley several flocks were noted, one consisting of several thousand individuals, all flying about three hundred feet above the ground on a southward course. In the nearby town of Gatlinburg, many were seen flying across streets and lawns, arousing considerable interest among the inhabitants. At Knoxville, Tennessee, several large roosting clusters were observed in trees at night. The strange thing about this flight was the fact that it was the first ever seen in this region, the usual flight path being down the east coast. One wonders what unusual conditions caused the butterflies to vary their migratory path. For those who are interested in butterfly migration, these flights occurred between October 1 and 5, 1970.

Like pale ghosts, Indian pipe flowers grow beside a stump. They have no green chlorophyll, obtaining their nourishment from underground fungi.

Beyond Huskey Branch the trail passed through an area of fallen timber, the results, no doubt, of a windstorm the previous year. I detoured to explore it and was at once rewarded by the sight of a large group of ghostlike Indian pipes (*Monotropa*). There were nearly a dozen growing out of the ground beside a log. Completely lacking in green chlorophyll, they were pale pink in color, with nodding blooms of the same shade. Among the strangest of all the plants of these mountains, they live as parasites, obtaining their nourishment from mycorrhizal fungi which, in turn, are parasitic on the roots of trees in a somewhat symbiotic relationship. Sometimes these peculiar plants are also called "ghostflowers" or "corpse flowers," both names being quite descriptive.

I had seated myself upon one of the downed trees—I did not notice what kind it was—and saw that it had lost most of its bark, revealing the weathered wood. What had attracted my attention was the fact that the grain spiraled up the large trunk, making two complete turns before reaching the first limb. This took my thoughts off on another tangent, not an unusual occurrence on my hikes.

Previously, I had seen a number of other dead trees, both in the valley and on the surrounding mountains, whose bark had fallen away, revealing spiral grain. Some of these trees had been standing, others fallen, but those with twisted grain had always aroused my curiosity. In the West I had seen many similar trees growing on high, windswept ridges. Some of those trees had showed grain with complete spirals in each twelve feet, and the odd thing was that most of them twisted toward the right, only a few turning in the opposite direction. Here in Hidden Valley most of the dead trees I had noticed spiraled to the right, bringing to mind an old fallacy. It is believed by some that trees in the Northern Hemisphere twist toward the right, while those in the Southern Hemisphere twist the other way. A similar fallacy is that vines north of the equator spiral in the opposite direction to those south of it. I had already settled this matter, in my own mind, by observing that about as many vines spiraled in one way as in the other.

In the case of trees, there is also this same inconsistency. While most of the trees I had noticed were "right-handed," there were many "left-handed" ones. Most of the maples spiraled toward the right, while the elms, if they twisted at all, spiraled to the left. Just what causes trees to grow in this fashion is unknown, but one thing I have noticed—a fact that others, too, have commented on—is that most trees growing in exposed situations, under adverse conditions, tend to develop twisted grain. However, the tendency to spiral seems to be partly an inherited characteristic. As an example, Scotch pines growing in northern Europe have straight grain, while the same pines in Germany have a corkscrew type of

Many trees have twisted grain, as does this dead tree still standing beside the trail. Why trees grow this way is not fully understood.

growth. When the seedlings of these trees are transplanted in the United States, they show the same peculiarities, proof that it is not, in this case, a matter of climate. Thus, I realized, that here was another unsolved, but intriguing, mystery of the forest. There seems no more logic to the direction of twisting of "barber-pole" trees than in the way in which vines twist when climbing up them.

While contemplating the spiral grain of the fallen log and speculating on corkscrew trees in general, I became aware of buzzing sounds emanating from a small maple about ten feet away. This seemed worthy of investigation, so I walked over and looked up into it. About ten feet above the ground, well hidden in the foliage, was a large hornet nest. It was nearly two feet tall and a foot in diameter, and was attached to a limb on one side.

Looking closely, I could just discern the nest's entrance, about

which buzzed a number of bald-faced hornets (*Vespa masculata*).
As I watched, workers were leaving and returning, presumably
on food-hunting forays. Even though some of the past nights had
been quite cold, the hornets were still busy, probably rearing
young. Somehow, I was saddened by the sight of this large nest at
summer's end, knowing that within a few weeks the thriving community would cease to function, its remarkable social organization
having come to a halt. Yet the hornets were toiling through the
autumn day, with business as usual, rearing more brood, building
more combs. What a waste of effort! Within a few weeks, fore-
warned in some fashion, many new queens and some drones would
be produced instead of workers. These would leave the nest and
mate; the drones would perish. In the meantime, each young
mated queen would locate a place to spend the winter, perhaps in
a hollow log, and settle down. These queens would be the hornets'
link between the summers. All the hot-tempered workers would

*By winter's arrival, this large paper hornet nest will be deserted, all the
bald-faced hornets dead. Only the mated queens will survive to found new
colonies in spring.*

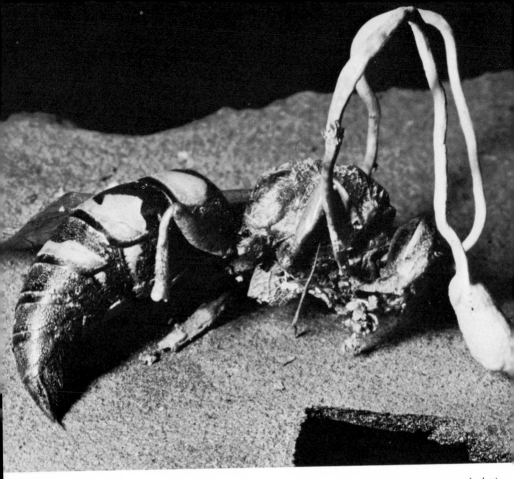

Many hornets become victims of Cordiceps *fungi that grow out of their bodies.*

have perished by the time winter arrived in the valley. Gales would destroy the carefully built nest, leaving only tatters of paper hanging from the bare limb of the maple. The hibernating queens, however, would be relatively secure in their winter quarters, their life-forces dormant, awaiting the arrival of spring when they would revive to found new colonies in the forest. Still, I knew that some of the hibernating queens would not survive the winter; some would be eaten by mice or shrews, others become victims of *Cordiceps* fungi. Several times I have found hornet queens that had been killed by this strange disease, the fungal sporangia extending up from their bodies on short stalks.

Large paper nests of the bald-faced hornets are quite common in the valley, as are the smaller nests of their relatives, the yellow jackets. Some yellow jackets nest in cavities in the ground, while others build small paper nests beneath overhanging ledges. In spring I have often found nests of the latter type.

While examining the hornet nest, I noticed the cast skin of a cicada still clinging to the trunk of the maple. The adult cicada had long-since departed from its outworn nymphal shell, had probably mated and laid its eggs in the twigs of a tree. After hatching, the nymphal cicadas had dropped to the ground and burrowed in, eventually attaching their sucking mouthparts to roots. They were probably now beneath my feet, busily pumping out root sap and slowly growing. When winter arrives and the trees become dormant, the nymphs will stop feeding, awaiting spring's arrival. Several years from now they will be full grown and emerge as winged adults to sing in the summer sun.

I have now walked the trail to Cucumber Gap at all seasons except winter, discovering the secret ways of its plant and animal life. Spring brings colorful flowers in amazing abundance, their bright hues contrasting with the vivid greens of new leaves. By summer the spring flowers have gone, replaced by others in lesser numbers. October brings cold nights and bright days bathed in radiant sun. The trees are then arrayed in splendid garments, their last farewell to summer. In time, they cast aside their leaves, standing stark against the mountainsides. The flowers have gone, replaced by seeds, the plants' hope chests for the future. Each animal, too, in its way, has made ready for winter's cold, waiting patiently for the time when the sun of another spring brings warmth to the mountains and the valley.

In several places on the ground grow earthstars (Geaster). When squeezed, clouds of spores are ejected from their spore sacs.

Chapter 8

A DAY IN MUSHROOM GLEN

AUTUMN IN THE MOUNTAINS, like spring, is a time of change. I do not know the valley in the winter and I have little desire to do so. Having been reared in the high mountains of the northwest, I can easily visualize the valley when it is blanketed with snow; I am sure that it must be very beautiful. Yet, like the local groundhogs and chipmunks that go to sleep in autumn and awaken in spring, I have no real awareness of the winter season here; I know only the halcyon months of warmth and sunshine, and am content that it remain so. Paralleling the habits of migratory birds that pass through the valley, I travel southward in late autumn, not returning until winter has gone.

Autumn is the time of colorful leaves and bright days—days that are often interspersed with periods of cloudy weather and rain. These wet periods, I have found, are favorable to the growth of fungi in almost fantastic variety. If the weather is warm after one of these autumn rains, fungi begin pushing up out of the earth or creeping over rotten logs in the shadowy forest.

Far up in Hidden Valley there nestles a glen. It is enclosed on one side by the foot of the mountain and by steep ridges rising on either side. Only on the side facing the stream is it accessible, and then only by crawling laboriously over fallen timber and pushing through a dense stand of hemlocks. It is a place visited by few, since it holds little attraction to anyone except a naturalist bent on exploration.

Here a mushroom, probably Lactarious, *pushes up through leaf-litter on the forest floor.*

The indigo lactarius (Lactarius indigo) has deep blue flesh and milky, bright blue juice. It is considered to be edible.

My discovery of this retreat was something of an accident. I had gone up the valley a few days after a rainy period, looking for native seeds about which I was in the process of writing a book. Along the way I had collected the maturing seeds and fruits of a number of plants and trees, including those of sweet shrub, silverbell, and strawberry bush. Along the river, the leaves of dogwood trees had turned scarlet, the tips of their branches adorned with bunches of bright fruit and, in one marshy area, I was gratified to find a large cluster of the glossy red "berries" of jack-in-the pulpit. As I moved by slow stages up the valley, I became increasingly aware of the abundance of mushrooms and other fungi, their growth stimulated, I presumed, by the recent rain. Thus, my attention was gradually diverted from fruits and seeds to fungi, and I soon was searching for these primitive forms of plant life to the exclusion of all else.

The past night had been warm and humid, conditions very favorable to fungal growth and, as I pushed on through the dense forest among the great pendant trunks of wild grapevines trailing up the trees, I inadvertently entered what became known to me as Mushroom Glen.

Forcing my way through the trees, I found my progress blocked by a large rotting log, its contours blanketed by a velvet-like growth of mosses, while a few ferns thrust their fronds up above the green surface. The log rested in deep shade, except for its central portion which was bathed in a beam of sunlight slanting down through the overhanging hemlocks. My eyes were at once attracted to a cluster of golden-yellow fungi rising through the mosses, spotlighted by the sun. These attractive fungi were so placed as to make it appear that they had been arranged for a miniature stage setting, lighted to show them off at their best.

I walked over and looked more closely at the fungi, deciding that they were a species of *Clavaria*, a type that often grows on decaying wood. This cluster was about six inches tall, the branched stalks or basidiocarps resembling marine corals except in coloration. I squatted down beside the moldering log to examine them

This bright yellow fungus (Clavaria) grew on a rotten log.

at close range, noting how their pale stalks branched and re-branched in graceful forms. Satisfied with my examination of the *Clavaria*, I climbed over the log and continued on my way, going around waist-high boulders whose tops were covered with matted growths of stonecrop and fern. Some distance ahead was another boulder, this one at least ten feet high. On its flat top I saw an animal of some sort. The boulder was bathed in sunlight, and the animal—I now saw that it was a large groundhog—lay indolently upon the stony surface, its body flattened against it. Even at the distance of fifty feet, I could see its beady eyes watching my every movement with wary concentration. There, in the sylvan glade we regarded each other, the groundhog and I. I was pleased at seeing the creature but, to it, I was an intruder to be carefully watched for signs of aggression. For nearly a minute it made no move, then it crawled down over the side of the boulder and entered its den beneath it. Unfortunately, my camera had been in its case and, since several minutes would have been required to change to a telephoto lens, the chance of an interesting picture had escaped me.

While examining the environs of the groundhog's den, I noticed a dead birch about six inches in diameter leaning against the boulder. Upon it I was surprised and elated at seeing a spreading growth of vivid blue-green fungi. So unusual was the color that I collected some specimens for later study, determining, eventually, that they were probably a species of *Cyphella*. Later, I saw other examples of this same fungi, always growing on dead birch trees or limbs.

Beyond the groundhog's domain I was forced to wade through masses of ferns and other vegetation as I approached the foot of the mountain, crossing a small brook that rippled over a sandy bottom between mossy banks where, in one place, I found another attractive cluster of coral fungi (*Clavaria*). This was, indeed, an ideal day for fungus hunting and so I continued on, now following up the stream.

About a hundred feet beyond the place where the coral fungi

Another coral fungus (Clavicornia), this one white in color and with branching stems, grew on a dead poplar.

had been discovered, I came to a point where there was a bare
patch of gravel beside the brook and upon which rested a dead
limb. Growing upon this limb was a large cluster of bird's-nest
fungi. (*Cyathus*). Each one consisted of a vaselike, flaring cone
with ribbed sides and a number of "eggs" resting within. They
did, in truth, remind me of tiny birds' nests, complete with eggs.
These "eggs" are actually spore cases, called *periodioles*, and it is the
manner of their dispersal by drops of rain falling into the fungal
cups that is of special interest. When a raindrop falls into one of
these cups, the periodioles are splashed away for some distance
where, presumably, they germinate and produce new crops of
bird's-nest fungi. Knowing the habits of these strange little plants,
I decided to take the time for a little experiment. I set up my
electronic flash and, from my finger, allowed drops of water
collected from the brook to fall into the flaring, "egg-filled" cups.
Sure enough, the periodioles were splashed away, some for distances up to a foot. As I dropped the water into the cups I attempted
to syncronize the flash so as to record the action on film. Later,
when the film was developed, I was gratified at finding that I had
been successful in capturing droplets of water splashing out of the
cups, carrying the periodioles along.

The method of spore dispersal used by the bird's-nest fungus is a
remarkable adaptation. Its cups are so shaped that their spore cases
will be splashed away to a maximum distance by falling rain. However, these peculiar little fungi obtain dispersal in yet another way.
When cows or other grazing animals accidentally eat the spore
cases, which are often attached to grasses, the spore cases eventually
fall to the ground, perhaps long distances away, uninjured by their
passage through the animals' digestive systems.

Attached to the side of a large tree near the brook I noticed a
bracket fungus of tremendous size, the largest I had ever seen. It
measured nearly eighteen inches across and its upper surface, deep
maroon in hue, appeared as if varnished. It was a *Polypore*. On this
expedition I found still other polypores, but none as large as the
first. I also discovered a number of large *Boletus* mushrooms grow-

Bird's-nest fungi (Cyathus) are among the most unusual of all fungi. Resting inside the cups are spore cases which are dispersed by drops of rain falling into them, as this high-speed photograph illustrates.

ing on the ground. One of these had a reddish cap and measured nearly a foot across. Its under surface and stem were yellow. I tentatively identified it as the bitter boletus (*Boletus fellus*).

Almost everywhere I looked some sort of fungus or mushroom was in evidence. There were beautiful Caesar's amanitas, brilliant red in color, some just pushing their unopened caps up through the leaf-litter. These mushrooms are closely related to the deadly amanitas, which are mostly white or creamy in hue. They, too, grew in several places among the tumbled stones.

The sun was now high in the sky, bathing the autumn forest in golden light. I walked on, climbing through tangles of wild grape-vines and dense growths of laurel. Suddenly, I noticed a number of large, dark-colored butterflies fluttering about a spot near a boulder some distance away. Cautiously approaching to obtain a better view, I could see that they apparently had been attracted to something on the ground that, at the moment, was hidden from view by tall ferns. The butterflies were red-spotted purples (*Basilarchia astyanax*) whose caterpillars, I knew, feed upon linden and other trees. In the West I had often seen a closely related species (*B. weidemeyeri*), clustered upon animal excreta on woodland trails. Thus, I naturally assumed that these butterflies had been attracted in a similar way, perhaps by bear droppings. This gave me some food for thought and so I paused, listening and looking in all directions. Convinced, at last, that there was no bear in the vicinity, I moved closer to the fluttering butterflies, discovering that, in addition to the basilarchias, there were a number of smaller butterflies, including some skippers. Some of them

Beside a boulder was a decaying mushroom which had attracted a number of red-spotted purple butterflies (Basilarchia). The butterflies had apparently been attracted by the odor of the decaying mushroom.

were perched on the rocky surface, while others crawled about over it in a most unusual fashion. Overcome with curiosity, I walked up to the spot to discover, if possible, what had enticed the insects. Much to my surprise, I saw that they had been attracted by a large mass of decaying fungus, probably a mushroom of some sort. What the fungus had been I was unable to determine. Later, I decided that it must have been some sort of stinkhorn, a peculiar mushroom having a vile odor attractive to flies.

The butterflies had been lured to the odoriferous mass of fungus, and were clustered around it like bees around honey. When I chased them away, they soon returned, as if attracted like a cat to catnip. Seemingly, they were not feeding upon the decomposing fungus, merely enticed to it, probably by its odor. It reminded me, a little, of a strange habit found in both wolves and coyotes, that of wallowing in and around the decaying bodies of dead animals. These carnivores apparently do not feed upon the carcasses; seemingly, they are merely attracted to them for some unknown reason. Ernest Thompson Seton, the naturalist, calls it the "carrion craze," but offers no explanation for it. As I watched the attractive butterflies fluttering about the decaying fungus I wondered if they, like the coyotes and wolves, were not afflicted by some sort of carrion craze, their chemoreceptors stimulated by the smell. I bent down and sniffed the fungal mass, but to my olfactory organs it was certainly not attractive. Why should butterflies normally enticed to flowers by their sweet-smelling perfumes, be so charmed by a vile odor? I must confess that I do not know.

Among the boulders scattered about the glen ferns grew in profusion, while upon the ground were many decaying logs, most of them covered with fungi of one sort or another. There were more branching coral fungi of golden hues, as well as clusters of small puffballs, the latter in the form of perfect spheres, looking like Ping-pong balls. Growing out of the side of a dead pine was an attractive little mushroom that I assumed to be a *Pholiota*. Its stem emerged from the bark and curved gracefully upward to its attachment beneath the cap. Almost everywhere I looked dead leaves

Growing out of the trunk of a tree was this mushroom (Pholiota) with its curved stem.

Looking like tiny parasols, these fairy mushrooms (Marasmius) grew upon a rotting log. They smell like garlic.

In some places were other tiny mushrooms (Mycena). They had grown on dead twigs and leaves.

and twigs had sprouted tiny mushrooms of graceful and delicate form. Small *Marasmius* mushrooms grew out of dead rhododendron leaves resting upon the damp earth or on the twigs of these same shrubs. Some of these miniature mushrooms had black stems, curving gracefully upward, topped by red, parasol-like caps. Clustered upon the decaying wood, they caused me to wonder if perhaps they belonged to forest fairies or gnomes. This brought to mind another thought: How unfortunate I was to live in an "enlightened" age when the belief in such fantasies is no longer tenable. How nice it would be to see gnomes, as did the ancient Greeks, guarding the treasures of the deep, dark forests! Such were my thoughts as I gazed at the delicate little mushrooms, but try as I might to endow them with magic, they remained miniature mushrooms growing out of the twigs and leaves, obtaining their nourishment from the decaying tissues.

Some of these tiny marasmius mushrooms are said to be edible, but I fail to see how enough could be collected for a meal. Some have garlic-like or peppery flavors, it is said, while others may be poisonous. Personally, I am content to admire them, leaving more intrepid souls to experiment.

Rejoining the brook, I discovered a dim trail following its course. This led, eventually, to the foot of the mountain where the water trickled down through a cool vale lush with mixed vegetation. Among the varied plant growth was a tangle of wild ginger, while beside a large boulder I noticed a group of most unusual mushrooms that I recognized as a species of *Hypomyces*. They were clublike in form and had no gills or pores, as do most mushrooms. Some distance away there was an older specimen, this one in its spore-shedding stage. When I prodded it, a cloud of brown spores drifted away from its top. The remarkable thing about these mushrooms—if you can really call them mushrooms—is that they are actually two different fungi living together, one parasitic upon the other. The parasitic fungus is an Ascomycete, more closely related to certain molds, though a number of large fleshy fungi, such as the delicious morel, belong to this same group. The host fungus may be one of several mushrooms, but when invaded by the parasitic fungus, its normal identity is lost. So altered in appearance does the mushroom become as to be unrecognizable. The fungus growth resulting from this strange combination may or may not be poisonous, depending upon whether the parasite has combined with an edible or a poisonous host mushroom. Among all the fungi that may be encountered in the forests, the *Hypomyces* is without a doubt the most unique.

Nearby, I was lucky enough to find some fine examples of the morel mushroom. Half-hidden beneath the dense fern growth, I noticed them by accident, the first time I had seen them in these mountains, though I knew they occurred here. These fungi, not really true mushrooms at all, are the gourmet's delight. Here on

Bottom right: *Among the most unusual of all mushrooms is this Hypomyces. It is actually a combination mushroom, consisting of two fungi living together.*

Above: *Found also in Mush-room Glen were poisonous am-anitas (Amanita). These mush-rooms are deadly.*

Right: *The bright red cap of a Caesar's amanita (Amanita cae-sarea) pushes out of the ground. This amanita is not considered poisonous.*

Left: *This is the edible morel (Morchella). It is fairly common in the valley.*

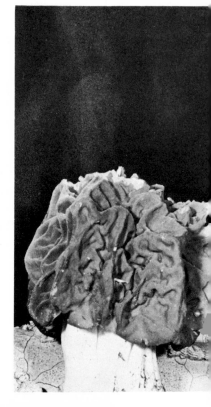

Right: *This is the false morel (Gyromitra). Unlike the true morel, it is poisonous. Notice the cloud of spores rising above it.*

Below: *This odd little fungus is known as the saddle fungus. It grows on the ground.*

the ground, beneath the ferns, their brown, honeycomb-like heads contrasted sharply with the green background. Some distance away was a specimen of the false morel (*Gyromitra*). Its cap was crumpled and dark in color. Unlike the morel, it is considered to be poisonous, although, strangely, some people may eat it with impunity. I touched this specimen and noticed that a cloud of spores drifted away in the air. There was another little mushroom in the vicinity that I had seen a few times. This was the saddle mushroom (*Helvella*), not too distantly related to the morel and the false morel. It has an offensive odor. However, British authorities state that it may be eaten.

Reference to the British brings up an interesting fact: probably because fungi are so easily dispersed by means of their microscopic spores, their distribution is almost world-wide. There is great similarity between the mushroom flora of almost every continent. European mushrooms are often identical to those found in the United States. I have a book on South African mushrooms and have been rather amazed at how many are the same as those growing here.

Needless to say, I was finding the day most rewarding as far as fungi were concerned. I had already found more in kinds and numbers than on any previous expedition in the valley, or, for that matter, anywhere else. It was time, I decided, to make my way back along the base of the ridge and down to the river. I paused now and then along the way to examine some plant or autumn flower that attracted my interest and, in the process, was forced to struggle through a clump of dog-hobble blocking the way. Beyond lay an open area covered, here and there, with club mosses and ferns, among which I was pleased to see several small mushrooms with bright scarlet caps. Their caps were cone-shaped and the tallest one reached perhaps six inches above the earth. These, I knew, to be a species of *Hygrophorus*, of which there are several kinds, most of them embellished with gay hues of scarlet and yellow. They looked almost as if they were made of wax, and were the most brightly colored mushrooms I had ever seen. In the vicinity there

Growing out of the side of a log was a Volvaria, a mushroom with a deep cup or volva at its base.

Below: Bracket fungi (Poly-porus) are common. These specimens show numerous growth rings or zones.

were several other mushrooms, some unusual and worthy of noting. One of these was the indigo lactarious, a kind I had seen previously but always found interesting because of the deep blue coloration of its tissues and abundant juice. With my belt knife I sliced into one of the present specimens and watched in fascination as the thick juice dripped from the cut like deep blue milk. The indigo lactarious is said to be edible, but I could never bring myself to eat such a strangely colored mushroom. One authority states that while it is probably nonpoisonous, it is acceptable only to the strongest stomach. My stomach is not that strong!

In many years of roaming forests in many locations, I have often found many fungi but never in the variety and abundance as on this autumn day in the valley of Little River. Evidently, on this first trip, conditions had been just right.

Reluctantly I turned my back on the little glen and walked down toward the river. I could hear its sound perhaps a hundred yards away. I had stopped to examine an interesting gall attached to the leaf of a white oak when, out of the corner of my eye, I caught a glimpse of something white in a nearby magnolia. Whatever it was had disappeared behind the trunk and I decided that I must have been mistaken. However, I continued to watch, mystified as to what it could have been. Perhaps, I decided, I had been deceived by light shining on a leaf. Then, suddenly, I was rewarded by a most unusual spectacle. A snow-white squirrel—obviously an albino —ran out on a limb and paused there in full view. Mushrooms were forgotten as I scrambled in my bag of photographic equipment for a telephoto lens. A minute or so was required to remove the close-up lens and replace it with the long-focus lens. In the meantime, the squirrel jumped across to a spreading rhododendron where I had an excellent view and opportunity for a picture.

This was the second albino animal I had seen in the Great Smoky Mountains; the other instance had occurred down in Cades Cove the previous autumn. My wife and I were driving down a wood-land road where there were many chipmunks, all busily engaged in gathering nuts and seeds and carrying them off to their dens. One chipmunk ran across the road, followed by another, the latter apparently carrying a piece of white paper. Looking more closely I found that it actually was a *white* chipmunk. It darted into a hollow stump and I at once got out my camera, supposing that it would eventually reappear. In this I was mistaken, so after nearly half an hour I gave up and walked over to investigate, discovering that there was a hidden burrow inside the stump, leading, no doubt, to the little animal's den. Deciding that the chipmunk was by that time curled up comfortably in its nest with no intention of reappearing, I left the scene. Later, I reported the incident to the Park naturalist's office and was informed that others, too, had seen the unusual chipmunk. A week or so later I returned, but saw no sign of the albino.

I have visited Mushroom Glen many times at various seasons, hoping to find fungi in abundance. The fern growth is always as luxuriant as ever, the brook still ripples along from pool to pool, and the mosses upon the surrounding rocks are green and velvety.

This cup fungus (Peziza) was three inches in diameter, brown inside and white outside.

An albino tree squirrel poses in a tree in Mushroom Glen. It was never seen again on later trips.

Everything looks the same as on my first visit, but I hunt in vain for abundant fungal growth. Here and there, of course, I do find a few specimens, but not many. Sometimes, too, I find myself looking for a snow-white squirrel in the dense foliage of the somber forest. Always I am mistaken. An albino does not survive for long; it is too easily spotted by enemies. Also, I remind myself, the mountains are vast and a squirrel is small; I doubt that I shall be privileged to see it again.

Chapter 9

USEFUL PLANTS AND DEADLY

THE PLANT LIFE of the valley is beautiful and varied; everywhere I look I see vegetation in astonishing diversity. Every plant is attractive in one way or another, yet I am well aware that among them there is danger, that death and beauty may go hand in hand. Some of the mushrooms of the valley are very colorful, yet within their tissues are contained highly toxic alkaloids that may bring violent death to the person so careless as to eat them. In the world there are more than 300,000 different kinds of plants, most of which are neither poisonous nor useful to man. On the other hand, nearly a thousand different plants are known to contain toxic substances and, strangely, some of these poisonous substances have medicinal value. Jimson weed found here in Little River Valley is very poisonous, causing delirium, distorted vision, coma, and even death. Yet tea brewed from its leaves or seeds has been used for the treatment of asthma, often, however, with grave results. Extracts from this plant were once used by medieval poisoners for the elimination of undersirable persons. But it contains an alkaloid known as *atropine*, a drug used in medicine to dilate the eyes and for other purposes. It also contains *hyoscyamine* and *hyoscine*. However, it is rarely

Jimson weeds (Datura) are highly poisonous, even though poultices of their leaves have often been used for the relief of pain or wounds. The dried leaves were also burned for the treatment of asthma.

eaten by animals because of its strong odor and bitter taste.

The term "alkaloid" is applied to a class of compounds synthesized by plants. The name was first given to these compounds because of their assumed alkali-like properties, yet many alkaloids are now known to have acid reactions. Chemists lump most of the toxic or medicinal substances found in plants in this rather broad category, chiefly because not much is as yet known about their true natures. Alkaloids come from plants, but not all plants produce them. Just why these, often toxic, substances are manufactured by plants is something of a mystery. We might perhaps reason that the plants derive some survival benefit from their presence, but chemists consider that they are, in truth, a sort of chemical by-product of the plants' metabolisms and that they neither help nor harm the plants in which they occur.

Quite common in the valley are squirrel corn and Dutchman's-breeches with their unusual little blooms. They are favorites of those who admire the local flora. Both of these plants are closely related to poppies and, like them, contain poisonous alkaloids that are chemically similar to those in the opium poppy. (Every kind of poppy contains alkaloids, but the opium poppy contains more than twenty different kinds.) Both squirrel corn and Dutchman's-breeches have caused poisoning of livestock. They are often called "stagger weeds" and, when eaten by cows, cause a staggering gait and loss of milk production. Other, more serious, symptoms may later occur. All parts of these attractive plants contain poisons.

Shakespeare was aware of the poisonous natures of some plants; he wrote: "Would curses kill, as doth the mandrake's groan, I would invent as bitter searching terms as curst, and harsh, and horrible to hear." The great bard was probably speaking of a member of the nightshade family found in Europe, but in the Great Smoky Mountains the term "mandrake" is applied to the common May apple, a plant that grows almost everywhere in the valley. All parts of the plant are poisonous, especially the roots, which contain a resin called *podophyllin* which has violent cathartic properties. It is said to have been used by the Indians for the treatment of various

Squirrel corn, closely related to poppy, contains poisonous alkaloids that cause sickness of cattle.

Below: May apple plants contain a resin called podophyllin, which has powerful cathartic properties.

disorders; perhaps it was a question of "fighting fire with fir
Recent research has revealed that substances in these plants ca
abnormalities in animal cell division and so have been investiga
with regard to the control of certain kinds of cancer.

On my explorations in the valley I have found many jack-in-t
pulpits, unusual looking and belonging to the arum family to wh
also belong calla lily and some other cultivated plants. Jack-in-t
pulpits are usually found along the bases of cliffs and in other da
places, but are frequently overlooked because their flower tubes
spathes are usually green. These plants contain poisonous oxa
crystals and this chemical is especially abundant in the tuber
roots or corms. These were eaten by the Indians, but were f
boiled to destroy the poison.

Pokeweed (*Phytolacca*) is a perennial herb frequently seen
the mountains. It grows taller than a man and bears drooping cl
ters of small, white flowers. Its deep red berries ripen in autu
and were once used in the dyeing of cloth. Frequently the you
shoots are used as cooked greens. These plants, however, cont
several toxic substances, including a saponin-like agent as well
an alkaloid called *phytolaccine*. Fortunately, cooking destroys
poison which otherwise may cause serious symptoms or even dea
It is believed to have some narcotic properties.

One of the most attractive spring flowers along Little River
dwarf iris, resembling closely the cultivated plant. As proof of
paradox that where there is beauty there may also be danger is
fact that the rootstocks of iris contain irritant substances capable
causing severe stomach upset. I was once called, late one night,
a physician who stated that a patient, a small boy, had eaten an
"root" and was quite ill. I assured the doctor that while the ro
of these plants do, indeed, contain toxic substances, their ingesti
is not usually fatal. It was not in this case.

Other members of the lily family, too, may be poisonous. One
these is fly poison (*Amianthium muscaetoxicum*), which I susp
may occur along Little River, though I have not seen it. Its ro
and leaves are poisonous to cattle and humans if eaten.

Jack-in-the-pulpits, especially their corms, contain poisonous oxalate crystals. The corms were eaten by the Indians, after cooking to destroy the poison.

In spring, violets of many kinds bloom in the valley, especially on the hillsides. White violets are often so abundant as to literally cover the slopes. One might consider violets to be the most innocent of plants, yet the roots or rhizomes of some have toxic properties.

A common shrub, often of tree size, in the valley is buckeye, with its spreading, palmate leaves. In autumn its nutlike seeds mature and drop to the ground. One fall, my wife collected a small basket of them and set it out-of-doors. Several hours later she noticed that most of her buckeyes had disappeared. Accused of giving them away, I denied the charge, explaining that I knew of no one who had need of buckeyes. Shortly, the mystery was solved; it was found that a chipmunk had been climbing up on the table and carrying them off, one by one, to its burrow beneath a stone. This surprised me, since I presumed these seeds to be quite poison-

Two nuts are usually present in each buckeye pod. It was once believed that one nut was poisonous and the other harmless.

ous and that danger lurked in both the nuts and the foliage. However, the diligent chipmunk was carrying them to its den where, I assumed, they were being added to its winter store of food. Did the little rodent eat them? If so, was it harmed? I do not know, but I am certain that if the nuts are harmful to chipmunks, they would long since have learned to avoid them.

While the eating of buckeyes is without doubt harmful to humans, it was once the custom of the mountain people to carry one in the pocket to ward off rheumatism, an ailment that was probably quite common in the cool, damp climate. Buckeyes have a long history of use in these mountains as warders-off of various ailments. An alcohol extract of the nuts was used for the treatment of hemorrhoids. Usually, the carrying of one in the pocket was believed sufficient to invoke its protective charm. It is said that fresh buckeyes are harmless, but that old buckeyes that have lain on the ground and fermented may harm hogs, cattle, or humans, resulting in inflammation of the mucous membranes and partial paralysis. However, it is believed that the Indians ate them with impunity after roasting. The Cherokees also employed these peculiar nuts in capturing fish; the nuts were ground into powder and placed in streams in much the same manner as I have seen natives of South Pacific islands using ground Barringtonia nuts. This stupefied the fish, which then floated to the surface.

Buckeye seed pods usually contain two seeds or nuts, and according to some "authorities" only one of each pair of nuts is poisonous. They say that squirrels and chipmunks know instinctively which one is edible; they eat the one and discard the other. This, of course, is absurd.

Other "cures" for rheumatism included teas concocted of burdock roots and seeds or from sassafras bark and roots. Rheumatism was also treated with teas made from colic root, spikenard, butterfly weed, sweet birch, black cohosh, and pokeweed. For external application, walnut bark, boiled and crushed, was believed to be efficacious.

Without doubt, the most common of all poisonous plants in the area is poison ivy, of which there are at least two kinds. The distinctive characteristic of all of them, however, is the presence of three leaflets, a fact known to most people who roam the forests of the Great Smoky Mountains. In spring, the fresh green leaves push up from the ground or open on climbing vines. Some poison ivy vines climb as high as fifty or more feet up the tall forest trees. The berries, when mature, are formed in clusters and are grayish-white in color. In autumn, the leaves turn scarlet, gold, or orange, adding additional color to the forests at that season.

The toxic agent in poison ivy is a phenolic substance called *urushiol*, which occurs in all parts of the plant from the roots to the fruit. Unfortunately, I am one of the very sensitive individuals and have frequently had inflamed hands and arms as a result of contact with it. My worst attacks have come from digging in the earth near the vines where I inadvertently had crushed the sap-filled roots.

Not all people are equally sensitive to poison ivy—or poison oak —since many can handle it with impunity. However, everyone should become familiar with the various kinds and thus learn to avoid contact with them.

The characteristic three-leaflet arrangement of poison ivy is known to most people, and many are sensitive to its poison.

A common white flower in autumn is snakeroot (*Eupatorium rugosum*). It belongs to the daisy family, grows from one to five feet tall, and bears clusters of small white flowers. Early settlers in the Smoky Mountains were often ravaged by epidemics of a fatal disease they called "milk sickness." The disease was believed to be contagious and, as a result, farms and even villages were often abandoned. It is now known that this "disease" results from the use of milk from cows that have grazed upon white snakeroot. This usually occurs in late summer when pastures are dry and green feed scarce. The leaves and stems contain *tremetol*, a poisonous alcohol. When eaten by cattle, various severe symptoms occur, including violent trembling, stiff joints, labored breathing and, possibly, death. Human symptoms include severe constipation, subnormal temperature, weakness, delirium, and collapse. This plant should not be confused with black snakeroot (*Cimicifuga racemosa*), also known as "bugbane" and "black cohosh." A tea made from this latter plant was once used for treatment of sore throat.

To flower lovers who visit the mountains in June and July the attractive blooms of rosebay rhododendron (*Rhododendron maximum*) are objects of admiration, and rightly so. Yet these beautiful shrubs with their dark green leaves and clusters of rose-colored blooms contain poisonous *andromedotoxin*, a resinous substance found in all plants of the family. Even the honey from the blooms of these plants is extremely poisonous. Xenophon, the Greek historian and soldier, tells in his *Anabasis* of a devastating illness that occurred among the ten thousand Greeks campaigning in Persia under Cyrus the Younger (401 B.C.). Apparently the malady was caused by ingesting plants containing *andromedotoxin*.

Another member of the poisonous heath family is mountain laurel (*Kalmia latifolia*), often called "poison laurel." This is a very beautiful flowering shrub found almost everywhere in the mountains where its attractive, rose-pink blooms are much admired. Symptoms of its effect on livestock, especially sheep, include irregular breathing, staggering, blindness, and even death. Since this shrub is very common, there has been conjecture as to whether or not deer in these mountains are poisoned by it. In Pennsylvania it

White snakeroot contains tremetol, *a poison causing "milk sickness" in humans who have used milk from cows that have grazed upon it. Many children have died as a result.*

was found that both mountain laurel and rhododendron may kill deer. A number of dead animals were found in forests where it was determined they had fed on the foliage of these shrubs.

Whether or not spotted cowbane (*Cicuta maculata*) occurs in the valley I do not know; I have not seen it there. It is, however, found in the general area. This poisonous plant, like the more common wild carrot or Queen Anne's lace (*Daucus carota*), belongs to the parsley family. They are quite similar in general appearance, but whereas Queen Anne's lace grows but a yard tall, the cowbane often reaches six feet and its lance-shaped leaflets are broader than those of the former. Cowbane is also known locally as water hemlock, though it is a different plant from the true water hemlock which grows in marshy places. Socrates, you may recall, was condemned to drink a brew of spotted hemlock, a common but very poisonous European weed that we now know contains *d-propylpiperidine*, a deadly alkaloid. The spotted cowbane of the southern mountains contains this same substance and it is a strange but fortunate fact that cows will not eat it—the reason for its name. The poison is chiefly concentrated in the rootstock and it is said that even a small piece of one will kill a cow. Symptoms develop rapidly, with convulsive spasms that become more violent until death ensues.

Another plant quite common in the southern mountains, especially in waste places, is cocklebur (*Xanthium echinatum*). I have seen a few specimens in the valley. An odd fact is that these plants are poisonous only during their early seedling stage when the first pair of leaves appear. Hogs and sheep have been killed by eating the plants at that time. The toxic substance is a glucoside called *xanthostrumarin*.

The most common poisonous snake in the valley is the copperhead. As much as I have roamed the area I have never been bitten, but I always wear boots, just in case. Incidents of bites by these snakes were once quite frequent, according to people who have spent their lives here. A few inhabitants still reside lower down the valley and, according to one, a most reliable person, his dog was

Cockleburs are armed with sharp, recurved spines. The plants are very poisonous during their seedling stage. Infusions of cocklebur plants have been used for the treatment of snakebite.

Left:Lily-of-the-valley contains digitalis, a powerful heart stimulant. It was once used as mountain medicine, but its use is dangerous, since the plants are considered to be poisonous. Right: The leaves and flowers of the mullein plant contain astringents that have been used for the treatment of coughs and for relief of pain.

recently bitten five times on the head by a copperhead. A strong brew was immediately made of cocklebur leaves and administered to the dog, which, by that time, was in a coma. Poultices of cocklebur leaves were also placed over the dog's wounds. Amazingly, the dog was up and walking around within an hour after treatment began and shortly appeared no worse for the experience.

At least one of the local ferns seems to have toxic qualities. This is the bracken or brake fern (*Pteridium aquilinum*), a common fern found in dry, sunny forest openings. While the poisonous material in these attractive ferns is unknown, it seems to be cumulative. In other words, the plants must be eaten by animals over a considerable period before symptoms of poisoning appear. Cattle develop high temperatures and there may be internal hemorrhaging. In some cases death may occur.

Many of the native plants of the mountains have therapeutic agents. The "wonder" drugs now in use are of relatively recent origin. Previous to their introduction, almost all medicines were compounded from plants, and many are still in common use, although I am sure that at least some of these home remedies were and are of little value except psychologically. In many cases, the patients survived in spite of the treatments, not because of them. In former years, physicians were scarce in the Smoky Mountain area, with the result that people were forced to rely upon the aid and knowledge of herb doctors, usually elderly women, who professed a knowledge of the medicinal value of the local flora. Many of these herb doctors were held in high esteem and their services were much sought after. I have been amazed, in talking to them and in reading about others, at the frequency in which the plants they prescribed were those that, in truth, do contain medicinal substances.

Several poisonous plants contain substances of value as heart stimulants, but it is a tribute to the skill of the herb doctors and to the vigor of their patients that they usually survived the treatments. Lily-of-the-valley and foxglove were found to be of value in heart stimulation. They contain *digitalis*, and both of these plants may be considered poisonous. Dogbane and American hemp are extremely poisonous, yet both plants contain cardioactive drugs. American hemp was used by the Indians in the treatment of dropsy.

Many other native plants containing useful alkaloids could be

Tea made from wild ginger roots has been used as a tonic and for the relief of gas on the stomach.

mentioned. It should be borne in mind, however, that infusions of some of these plants, taken internally in sufficient quantities, may cause grave illness or even death; thus, their use should not be experimented with by an amateur. It is a fact, however, that the early settlers and the Indians before them did make use of many plants that we now know to be quite poisonous or at least suspect.

The native plant remedies used by the mountain people were as varied as the ills to which they were heir. For the treatment of coughs, teas made from the following plants were used: wild cherry bark, goldenrod, persimmon bark, black cohosh root, pine needles, sarsaparrilla, boneset, sassafras bark, mullein, sage, mint, linden leaves, sweetgum bark, and root of maidenhair fern. Poultices made from the blossoms of Jimson weed were applied to wounds for the relief of pain.

Digestive disorders were treated by teas concocted from wild ginger roots, colic root, blackberries, snakeroot, burdock, sourwood leaves, boneset, ginseng root, holly berries, Dutchman's-pipe, and butternut bark.

As purgatives and laxatives, the following infusions were used: sycamore bark, holly berries, butternut bark, iris root, May apple root, dogbane, fairy wand, wahoo, white ash bark, pokeweed, and Indian pink. As a bulk laxative, the seeds of plantain were soaked in water, causing a clear gum to exude. This gum has also been used as a hair-set. For the cure of dysentery, teas made of pigweed, oak bark, and geranium were favorites. Oak bark contains tannin, a strong astringent. Pigweed was also recommended for ulcers and hemorrhage. For stomach gas, infusions of yarrow and wild ginger were used.

There must have been great need for tonics, since a wide variety of plant teas were used, including those made from maidenhair fern root, fairy wand, spotted wintergreen, Joe-Pye weed, white ash bark, hydrangea, senica-snakeroot, Solomon's-seal, willow bark, foamflower, prickly ash bark, and black cohosh. (The name Joe-Pye weed, by the way, comes from an early New England Indian herb doctor by the name of Joe Pye, who is supposed to have cured typhus fever with this plant.)

Roots of Indian pink were once boiled and the tea used for the treatment of intestinal worm infections. It was also used as a cathartic.

A tonic was made from prickly ash bark. Tea made from the bark caused sweating.

Emetics were concocted from several plants such as butterfly weed, boneset, pokeweed, senica-snakeroot, Solomon's-seal, and wake-robin trillium. For elimination of internal worms, infusions made from the bark of mountain maple, butternut, ambrosia (*Chenopodium*), bee balm, Indian pink, and oil from the oil nut were employed. Oil of chenopodium was a standard remedy for internal parasites in both men and animals.

Bloodroot, a common and attractive plant, is extremely poisonous. However, it has been used for the relief of pain and as a sedative. Another poisonous plant is horse nettle, yet it was used for the treatment of convulsions and epilepsy. Infusions of Hercules'-club were also believed to be of value in cases of convulsions. Less dangerous were poultices made of various native plants, some of which may have had value in the reduction of swelling or for the relief of pain. A few examples are Jimson weed blooms, witch hazel leaves, and prickly ash bark. Dock leaves were applied externally for snake bite or for boils. Salves to hasten the healing of wounds were made from sweetgum roots, St. John's-wort, chickweed, plantain, dock, hickory bark, and dandelion leaves. For the elimination of moles and warts, the white sap of milkweed was recommended. A tea made of lady's-slipper roots was used for the relief of headache, while the smoking of dried henbane leaves in a pipe was thought to have value in cases of "nerves."

One of the most peculiar plants found in the valley is the Indian pipe, often known as ghostflower. The Indians once used a lotion made from these plants for the strengthening of the eyes. Solomon's-seal is a common and attractive lily in the valley where it grows in moist locations. Regarding it, an early European herbalist

The roots of bloodroot are poisonous, containing morphine. However, the plant has been used as a laxative and for other medical purposes.

Below: *The bark of Hercules'-club also has been used to induce sweating.*

The white sap of the milkweed was believed to be a cure for warts, while the roots were used in making teas for the treatment of rheumatism.

The horse nettle is extremely poisonous, but was once used for the treatment of epilepsy. It belongs to the nightshade family.

named Gerarde states that if the root is applied when "fresh and green, taketh away in one night, or two at the most, any bruises, black or blue spots, gotten by falls, or women's wilfulness in stumbling upon their husband's fists."

Miscellaneous ailments for which plant remedies were used included infusions of bedstraw (*Galium*) for kidney infections, curly dock for ringworm (a fungus infection), and sweet gum for scabies. The twigs of sweet gum were chewed to clean the teeth. In some cases, these twigs were soaked in brandy before use, a practice that modern makers of toothpaste might take cognizance of!

Honey was a favorite therapeutic agent with many people and had a wide variety of uses, including treatment of burns, hayfever, coughs, and "nerves." Too, honey, along with maple sugar obtained from local sugar maple trees, was the usual sweet. Beer was made from birch.

A common practice in spring was the taking of "tonics" or "bitters" in the belief that they thinned the blood and so were beneficial. Such teas were made of cinquefoil, dandelion, burdock, wild ginger, and sassafras. One early traveler, when asked if he would like a drink of "bitters" to thin his blood, replied, "My blood is so thin now that I can hardly walk."

Perhaps the most remarkable drug plant of the Great Smoky Mountains was, and is, ginseng root, obtained from the ginseng plant (*Panax quinquefolium*). It belongs to the plant family Araliaceae to which also belong English ivy, (*Hedra helix*), wild sarsaparilla (*Aralia nudicaulis*), and Hercules'-club (*Aralia spinosa*). Ginseng, often called "seng" or "sang" by the mountaineers, is not an unusual looking plant. It grows a foot or so tall and has five ovate leaflets, palmately arranged. Found only in rich, shady woods, it grows best where there is considerable moisture. In autumn it produces a cluster of bright red berries and soon loses its leaves. The root, when mature, is ovate in form.

Shortly after the first settlers came to the mountains, they began gathering ginseng roots, mostly for export to China. The roots were also used locally for almost everything from "nerves" to asthma,

Ginseng berries are produced in autumn and are red in color. The tubers or roots of ginseng have been used in China for many centuries as tonics and love potions. Many tons of ginseng root are exported each year, most of it grown commercially.

and they were thought to have value in the cure of tuberculosis. In Oriental lands ginseng root is chewed in the belief that it imparts strength, especially sexual vigor. It is alleged, too, that the root has the force of "heavenly fire," by which health and happiness are assured. In China, ginseng is used in the making of love potions. In Russia, scientists are reported to have found various compounds in the root with therapeutic value. By contrast, *The Dispensary of the United States* states that ginseng root has no real medical value. Yet, it is still gathered in the wild or grown commercially for export, at present bringing about fifty dollars a pound. Considering this price, it is little wonder that the rangers of Great Smoky Mountain National Park must be continually on the watch for "sang" poachers. The truth is that it is now a very rare wild plant in these mountains.

Among the early "sang" gatherers in this general area was Daniel Boone, who took it up the Ohio River for shipment to Philadelphia where it brought thirty-four cents a pound at that time. George Washington notes in his diary that, "In passing over the mountains I met numbers of persons and pack horses going in with ginseng."

At present, most of the "sang" root exported from the United

States is grown commercially. However, the cultivated dried ginseng root brings only about ten dollars per pound as compared to about fifty dollars per pound of wild root. The seed is sown in beds during the fall, winter, or early spring, and the growing plants must be protected from the sun by arbors of cedar branches. About four years are required for maturity, at which time each root weighs about three-fourths of an ounce. At this rate, an acre produces up to 5,000 pounds of "sang" roots that are harvested in autumn. Ginseng growers, of which there are a number, also sell the seed, at about twelve dollars a pound.

In addition to their medical uses, plants and plant materials were once employed in the dyeing of yarns and cloth. Baskets made of native materials, such as honeysuckle vines, were also dyed with plant infusions. Such baskets may still be purchased on the Cherokee Indian reservation located just south of the Great Smoky Mountain National Park. Some of the plants used and the colors imparted were: alder bark—brown; cocklebur—chartreuse; curly dock roots—yellow; mountain maple bark—rose-tan; jewelweed—yellow; Joe-Pye weed flowers—red or pink; pokeberries—red; rhododendron leaves—gray; sourwood bark plus walnut hulls—black; sumac berries—dark gray; tulip poplar leaves—yellow.

Ginseng plants produce tuberous roots which are collected and dried for the market. In George Washington's day these dried roots brought only a few cents a pound. Present prices range as high as fifty dollars a pound.

The Great Smoky Mountains are very old, far more ancient than the Rockies. They rise more than 6,000 feet, dissected here and there by deep, forest-filled valleys.

Chapter 10

IN THE BEGINNING...

THIS MORNING, in a pensive mood, I walk up the valley, following the river and listening to its music. It is May, and here and there the river bank is starred with flowers, trembling now and then in the morning breeze. Towering high above the forest floor rise great sycamores and tulip poplars, their trunks bathed in bright shafts of sunlight. Farther skyward, against a backdrop of fleecy clouds, a red-tailed hawk coasts gracefully on quiet wings toward the ascending sun.

The valley is peaceful and I am content to drink in its beauty and quietude. This, I pause to consider, is a place where things do not change, where yesterday, today, and tomorrow are one and the same. Yet, even as these thoughts course through my mind, I suddenly realize that these hills are not eternal at all, that even as my feet plod up the quiet valley, changes are slowly taking place around me. Hour by hour the boulders of the river are being worn away; the bed of the stream is being lowered, and this morning it is a little lower than yesterday. Gradually, these very mountains are being washed away; eventually they will be gone and the place where I stand will be a level peneplain. It has happened before, and will probably happen again. In Nature's continuing sequence, mountains rise only to be ground down once more by rushing waters. These mountains will, in time, disappear as the world spins on through space and time, but following the recurrent cycle of mountain building and great tectonic forces, they will perhaps rise again from their "ashes," inhabited by new kinds of birds singing from trees not yet evolved.

Such thoughts are perhaps a little disturbing as I meander along the river on this pleasant morning. Yet, even in change there is a kind of permanence, a knowing that time will go on and that the cycle will continue into the far distant future, a time too remote for the puny mind of man to visualize. As it was in the past, so will it be again and I must be content to enjoy the here and the now, knowing that enjoyment at best is a fleeting thing. I must appreciate the beauty of this valley, conscious that most of its long story lies moldering in the past.

In such a thoughtful frame of mind I seat myself upon a mossy log. A catbird calls from a clump of blooming dog-hobble, while a Carolina chickadee swings its small, slate-colored body beneath a sourwood twig. Only half-conscious of my surroundings, the present fades away and my mind slips back to Pre-Cambrian time, that period of earth history more than half a billion years ago. In my fancy, I see the archaic landscape stretching away, grim, forbidding, and devoid of life. Far to the east I see a great mountain range, blue in the distance and with jagged peaks. Faintly I hear the thundering sounds of a volcano and see its cone rising from the plain. From its top there drifts a column of white smoke, and rivers of molten lava flow slowly down its side. But there is no other sound, no song of bird nor chirp of cricket, and never a clump of green to ameliorate the harshness of the scene.

Millions of years seem to pass, and when I look again, the world has entered a new period in the time scale of earth history; Pre-Cambrian time has gone and the Cambrian has dawned in the continuing drama of change. Toward the west there now stretches an inland sea and I see white breakers thundering upon an ancient shore. But the land is still devoid of life and, except for the distant sea, I could be upon the surface of the moon. Out there in the sea I know that there are now simple forms of life, but they are not evident and will leave only fossilized remains as proof that they ever lived. There are creeping trilobites, shelled brachiopods, and a few snails, sponges, and worms. Primitive marine animals are just now coming to vogue, joining the simple algae and bacteria that probably had their origins in Pre-Cambrian seas. Upon the land, in sheltered locations, some algae have now divorced themselves from the water and become fungi, primitive, creeping forms of plant life. Almost three-fourth's of earth's history has already passed and, while a few living things have appeared, mostly in the sea, there is little enough to show for the vast period of time that has elapsed.

More time passes, and the scene changes. The mountains toward the east are no longer abrupt and rugged; there is falling rain, and sparkling streams rushing down the slopes, cutting away the rocks and carrying the debris down to the lower elevation. The streams flow across the plain, finding their way, at last, into an inland sea, slowly filling it with sediments eroded away from the distant range . . .

From my readings of historical geology of the Great Smoky Mountain region, I know it to be the belief of most authorities that during the Pre-Cambrian period there existed a great mountain range to the east of the present Smokies. Just how high these ancient, unknown mountains were can only be inferred from the large amounts of material that accumulated on the western lowlands and in the inland sea from their wearing away. This range, known to geologists as Appalachia, was originally very high, even higher perhaps than the Alps or the Tetons of Wyoming. Eventually, the earth-borne forces that had elevated them gradually subsided, and streams rushing down their precipitous slopes began cutting them away. As millions of years passed, the ancient stone was gradually eroded. The streams on the eastern, or Atlantic, side entered the sea and dropped their sediments in deep water. Perhaps, too, part of ancient Appalachia sank beneath the sea. Streams on the western slopes flowed down across a plain and into an inland sea extending all the way from Canada to the present location of Alabama. This narrow, inland sea lay west of the present moun-

In Newfound Gap (elevation 5,048 feet) are exposed strata of shale formed half a billion years ago near sea level by the wearing away of an ancient range of mountains toward the east called Appalachia.

tains, having been formed by the sinking of the land, creating a trench or sea trough known as a geosyncline. This waterway was connected with the sea at both ends, but its actual width is unknown.

As time passed, the streams rushing down the western slopes of ancient Appalachia cut them away, leaving deep canyons surmounted, no doubt, by rocky cliffs. It was a dreary landscape; no forests clothed the mountainsides to slow down the processes of erosion. Land vegetation was far in the future, and so the falling rain gathered into torrents that hurried down the mountains like silver ribbons, eroding them very rapidly.

That the streams of ancient Appalachia were very swift we have ample evidence in the kinds of materials they carried. The sediments they deposited on the plain and in the inland sea tell us that at first the mountains were steep, since the rock debris was very coarse, having of necessity been carried by swift currents. These sediments consisted of gravel and coarse rock fragments.

As the mountains were gradually worn down and became less

steep, the flow of the streams decreased in rapidity, transporting finer and finer materials, and depositing them, layer upon layer, at the foot of the mountains and across the plain. Thus, in effect, the streams flowing down the slopes of ancient Appalachia wrote their own history. In time, the mountains were gone, cut down, bit by bit, by the water that fell upon them. The debris was spread upon the plain and in the sea in successive layers or strata, the most recent upon the top. As millions of years passed, the rock fragments, sand and gravel, hardened into sandstones and conglomerates. Gradually, the inland sea, no longer muddied by the stream-borne sediments, became clear, a fit habitat for primitive marine life of many kinds. These living organisms thrived and died in the ancient sea, leaving their outworn, calcareous shells to settle to the bottom where they accumulated in vast numbers, forming thick beds of lime. Slowly these beds hardened into limestones, resting upon the older strata of sandstones and conglomerates that had been laid down during Pre-Cambrian time. Much of this ancient limestone now lies be-

In Hidden Valley, the author stands beside an exposed portion of the stony skeleton of the mountains. Now partly covered with mosses and lichens, this stone was formed before living things appeared.

neath portions of the present Great Smoky Mountains.

Thus, where once had stretched the great mountain range of ancient Appalachia there was now a level peneplain; the old mountains were gone forever. Toward the west, where once had rolled an inland sea, there was now dry land, a scene of dreary desolation, still devoid of life.

The sediments that had been washed down from ancient Appalachia and spread out upon the plain and in the inland sea had, as we have seen, hardened into stone containing fragments of quartz and feldspar. Ever so slowly these elements of mountain debris had been metamorphosed into stone and lay in layers or strata upon the earth's surface.

Thus came to an end the first portion of earth history, the long period of rock making. For the most part it had been a quiet time. Yet, during its span, great mountains had heaved upward in many parts of the world. A great flow of lava had gushed from earth's interior and flowed out over Canada, covering two million square miles, remaining today as the Canadian Shield. Volcanoes had sometimes been active in the region where the Smoky Mountains now rise. They had poured molten material into cracks in the ancient sedimentary stone, leaving *intrusions* of quartz that today may be seen as narrow white bands in boulders tumbled down from the mountains. In time, the volcanoes had ceased their outpourings, never again to be active in the region. Their role in the formation of the mountains had ended.

New theories of ancient geology have added another possibility to the origin of the strata that make up the present mountains. There now seems abundant evidence that the world's continents have not always been located where they are today. According to this theory, the present continents originated by the breaking up of a large, original land mass called Pangaea and moved slowly apart until they were situated in their present locations on the globe. If you will look at a map of the world, you will see that if North and South America were to be moved across the Atlantic Ocean they would fit rather well against Africa and Europe. Thus, according to this new concept of continental origin and drift, the strata making up the present Great Smoky Mountains may have originated from sediments washed down from ancient mountains located in France. This may seem farfetched, yet there is accumulating evidence that something of the sort did occur.

In most cases, mountains rise from depressions in the earth's crust called geosynclines, or sea troughs, and the birth of the Great Smoky Mountains was no exception. The weathering away of the mountains of ancient Appalachia had spread sediments across the plain to westward, covering it and filling the sea trough with beds perhaps 40,000 feet in thickness. But within the earth great forces were commencing to stir. Strange currents deep below the earth's crust began sluggish movements. Ever so gradually the plain to the west of old Appalachia began to heave and buckle. The surface of the land crumpled and wrinkled under the power of gigantic forces that folded and compressed the horizontal rock strata into fantastic forms. The time of rock formation was over, and the stage was set for the next chapter in the Great Smoky Mountain story. The period of active mountain building was at last under way. The Paleozoic Era had dawned, bringing with it living things in great diversity, but the dry land was still devoid of life. The time was 400 million years ago, a period of momentous change. The new Appalachians would be pushed upward as gigantic peaks, towering perhaps five miles into the skies.

Slowly, ever so slowly, great ridges were crumpled upward, some of them folding completely over in looplike flexures. Those strata were gradually compressed by earth's restless forces. As time passed, the strata of Pre-Cambrian stone were slowly pushed northwestward, sliding over the top of the more recent limestones. It was like a rug being pushed against a wall. As the tectonic forces, irresistible in extent, pushed the strata toward the northwest, they rose, buckling upward in gigantic folds and loops. During this process the older strata were thrust slowly over the newer strata as they slid across the land. Thus, in effect, the Great Smoky Mountains have been moved many miles toward the northwest.

Here, the ancient strata are strongly tilted, the result of the vast tectonic or rock-bending forces that formed the mountains.

While most of the rocky skeleton of these great mountains are of Pre-Cambrian origin, they now rest upon sedimentary beds laid down at a much later date. As a result, the strata have been reversed, so that the older rocks now rest above the newer ones. This momentous occurrence is known as the Great Smoky Overthrust. The leading edge of the old Pre-Cambrian strata lies today in the vicinity of Chilhowee Mountain, along the lower reaches of Little River. And so, most of the rocks that make up the Great Smoky Mountains, tremendous in mass and extent, have been moved from their original place of formation. In the process of compression, the width of the mountains was decreased to about half its original size. It was as if a giant hand had pressed against them, shoving them northwestward. Just how far they were pushed no one knows; perhaps they were moved as far as a hundred miles.

Since their creation these massive mountains have, of course, been reshaped by the erosive action of rushing streams that cut deep valleys and gave them their characteristic conformation. On the far eastern slopes the original highlands were eroded quite rapidly. There the stone core of the mountains was composed mostly of red sandstones and shales which were cut away much faster, leaving a wide piedmont stretching to the shores of the Atlantic Ocean. Nearer the mountains broad valleys were eroded. On the western slopes, however, the erosive processes were much slower; instead of broad valleys worn down by streams meandering back and forth across them, the streams, such as Little River, sliced slowly downward through the harder stones, cutting narrow gorges, often with vertical walls. Some of these gorges are now several thousand feet deep and impart to the western approaches to the mountains their rugged grandeur.

In many parts of the mountains there are other evidences of the great, slow-acting forces that created them. Not only have the strata been folded and crumpled, but in the process, breaks or faults

This large boulder, worn smooth and rounded by millions of years of rushing water, shows evidence of ancient volcanic activity. It was split and the crack later filled with lighter-colored igneous stone.

Boulders as large as a cottage rest in the river, their contours rounded by the abrasive action of the flowing stream. On top of this one are fragments of crayfish, all that remains of a raccoon's meal.

have occurred. It was as if the mountains had been sliced here and there by a great knife and the severed portions moved with reference to each other. One of these, the Oconaluftee fault, stretches along the foothills on the southeastern side of Cades Cove. It appears as a break in the slope. Another fault is the Greenbrier fault, which may be seen when you look eastward from Maloney Point at Fighting Creek Gap. This fault bisects the lower slopes of Mount LeConte and continues on toward the northeast.

In several places in the mountains there are basin-like hollows called *coves*—sequestered little valleys where cattle now graze in peace and where deer emerge from the surrounding forests at dusk to feed on the tender grasses. But these pleasant coves did not just happen; they came into being as a result of the same forces that carved the mountains. The ancient strata, you will recall, were pushed over the newer limestones, folding and bending as they moved, thus creating the great mountainous mass we cell the Great Smoky Mountains. These strata, being very old, contain no fossils. As the mountains were cut away by streams, the limestone was, in places, exposed. Since this exposed limestone was softer than the ancient sandstones and conglomerates above it, the limestone was eroded away more rapidly, forming the hollows we know as coves. Thus, the floors of these coves consist of limestone, the reason that it is known as "valley" limestone. The coves are, in effect, windows in the thrust sheet of older stone through which the newer limestone beds may be seen. Since these beds of limestone were formed at a later time than the strata above them, having been laid down in the inland sea, they contain abundant fossils of ancient marine organisms.

Eventually, there dawned the Triassic period which began about 250 million years ago. During this long span of time the mountains underwent prolonged erosion. Deep valleys and gorges were carved by the rushing streams and, at last, the mountains appeared in approximately their present form. By this time they were clothed in luxuriant vegetation. About a million years ago, for reasons that we as yet do not understantd, world climates became cooler. The Pleistocene with its Ice Ages was approaching. Thick layers of ice

A view of Cades Cove. The floors of the coves are formed of limestone, exposed by the wearing away of the ancient Pre-Cambrian strata that once covered them. Cattle and deer now feed in these grassy valleys.

The limestone of the coves contains many fossils of ancient sea animals such as trilobites.

crept slowly downward from the pole and lay a mile or more deep over Canada and much of northern United States. Glacial ice ground down across the land and through the valleys, reshaping them and changing their topography.

The Great Smoky Mountains were never actually visited by glacial ice, but the great mass of ice to the north breathed out its frigid breath and the climate of the mountains turned cooler. The tops of the mountains, which once had been covered with dense forests, then became too cold for forest growth and soon all but the most hardy forms of plant life disappeared. The mountaintops then lay above timber line and, possibly, were buried under snow during much of the year. In the cooler climate, plants and trees that had lived much farther north now extended their ranges southward. The changing climate also stimulated plant evolution, with the result that the already diversified vegetation of the region became even more varied.

What we know as the Ice Ages was not, as some suppose, one

long, continuous period of cold; it lasted for perhaps half a million years, interrupted periodically by warmer times during which the ice receded northward. After each warm period, lasting probably several thousand years, the ice sheet moved southward, and once more the vegetation of the mountains was subjected to colder weather. Tempered and stimulated by these changing climatic factors, the plants of the Great Smoky Mountain area flourished and changed, becoming even better adapted to various habitats. As a result, this mountain area emerged as a great natural laboratory for the development of plant life. In time, these modified plants spread out over almost all of eastern United States. In a manner of speaking, these mountains were the cradle of eastern plant life.

Eventually, the Ice Ages came to a close; the thick layer of glacial ice disappeared from all but the highest mountains in the United States, and the stage was set for the next chapter in the Smoky Mountain story. The last cold period ended about 20,000 years ago, and since that time the climate has probably remained about the same as at present. The Ice Ages are gone, but they left their imprint on the mountains in several ways. Trees of more northern

Pieces of shale, tumbled about for millions of years in the river, are now smooth and rounded.

areas, such as spruce and fir, now grow at the higher elevations, having become adapted to life there while the climate was cooler. These form dark forests on the steep slopes above 5,000 feet elevation. Hemlocks grow lower down; they thrive along Little River, their feathery boughs dark against the mottled trunks of the sycamores and birches. It is estimated that more than half of the trees now found in these mountains are northern species.

On the higher Rocky Mountains there are true timber lines, definite elevations above which trees do not grow because of climatic restrictions. By contrast, there are, at present, no true timber lines on the tops of the Great Smoky Mountains. However, many of the higher elevations are at present bare of trees, covered only with growths of shrubs, grasses, and ferns, many of which are characteristic of areas farther north. These are known as "balds," and often remind one of lowland meadows. A few of these "balds" are accessible by automobile. I once drove up to a "bald" known as Stratton Meadows located farther south in the Cherokee National Forest. The road is very narrow, a mere forest trail. The word "meadow" was, I found, very descriptive. The area was covered with tall grass with, here and there, copses of small trees and shrubs. A short distance away a herd of cattle grazed peacefully and I heard the sound of their bells. This led me to the conclusion that these were milk cows and that there should be a farmhouse nearby. Later, I learned that no one lives near Stratton Meadows; the cattle I saw were being pastured there during the summer, the bells used to keep the herd together. It was a pastoral scene, and I found it difficult to believe that I was high on the mountains along the boundary between Tennessee and North Carolina.

In reality, there are two types of balds, grassy balds and heath balds, the latter often called "laurel slicks." Grassy balds are open, meadow-like areas, with shallow soil and scattered shrubs. Examples are Silers Bald, Parsons Bald, and Andrews Bald. The soil of laurel slicks, or heath balds, is characterized by the presence of moist peat, often to depths of two or more feet. Here grow dense covers of rhododendron, mountain laurel, and other shrubs of the heath family. Examples are Madrons Bald and Myrtle Point, the latter on Mount LeConte high above the headwaters of Little River.

Whatever their origin, "balds" are a unique feature of these mountains and of interest to both plant ecologists and laymen alike. To the botanist, they pose interesting problems of plant geography and ecology; to the hiker, they afford superb views of the vast mountain panoramas stretching away below. Without the advantage of these open, forestless areas, the view of the traveler is usually restricted to his immediate vicinity by the dense, jungle-like walls of surrounding forest.

Because these mountains are very old, with most of their story in the long past, my thoughts are often retrospective while exploring Hidden Valley. Everywhere I look there are evidences of bygone days. The topography of the mountains shows evidence of time's slow sculpturing, of the carving action of the streams. The present is the product of the past. Even the sound of Little River has a timeless quality; I often awaken at night, its soft murmur penetrating my consciousness with a tone that varies only in volume. It was the stream that carved the valley; it is the stream that carves it still.

Truly, these are venerable mountains sleeping gracefully in old age. Their tumultuous youth is gone, their once-jagged peaks worn down and their slopes now carpeted with lush vegetation. They rise, tier after tier, into the smoke-blue haze breathed out of their deep, dark valleys. Today, they rest upon the ancient land, massive monuments to an eventful past.

Hidden Valley is but one small part of these great mountains, yet it is representative of the whole. Each time I visit the valley I am reminded of its past by the sight of exposed strata at various points along the river. Often I pause to examine these folded and tilted layers of archaic stone, encrusted now with lichens and mosses. I recall that this is an ancient valley, but I am grateful for the privilege of enjoying it now. In spring, I marvel at the colorful wild flowers and watch as the trees slowly unfold their lacy foliage. In autumn, I gaze entranced at the slopes splashed with multihued glory, and it is with a feeling of nostalgia that I see the leaves drop, one by one, decking the forest floor with a colorful carpet over which my plodding feet make rustling sounds. Heard always are the soft background tones of the streams and the waterfalls, of the wind in the trees, and the sonorous voices of insects. This is Hidden Valley.

Index

Page numbers in *italics* are those on which illustrations appear